Source Power
—New Trend Office

源动力
——新锐办公空间

未来文化 主编

中国·武汉

PREFACE 序言

引言：

每一次因极端气候带来的惊悚新闻，已使得人们不得不正视环保的迫切性，而这样的趋势，也让设法减缓环境劣化的种种概念、作为，开始从日常生活扩展到各种空间、建筑、生态，也因此衍生出「绿能智慧」这样的新名词。

内文：

身为专业设计师，我对环保的认知是非常直接的一种反射领悟，当然在个人或家庭这类的社会基础单元中，要落实环保精神并不难，不过就多人共享的大型厂办或商空来说，智慧统合与节能减碳既然是未来建筑的趋势，相关环保节能的议题，就需要更多的专业与科技方能有效执行，例如针对用电效率、用电安全、光能+人工照明、植栽、水循环利用、冷房效益等等，都是大型厂办商空优先的节能重点，其中有关用电的部分，可以透过需求管理和用电监测，让业主轻松控制营运成本。

当节能智慧建筑已经成为未来市场主流，世界各国也不遗余力加以推动，好比座落于西班牙-巴塞隆纳，由当地政府协同知名建筑师Enric ruiz-geli合力打造的Media-TIC大楼。这座大楼除了是新颖的创意养成中心，也是建筑节能环保的模范，科技感十足的外观由铁氟龙涂料板组织而成，与北京奥运场馆-水立方所使用的相同材质，可有效阻绝紫外线，发挥绝佳隔热效果；而在建筑内部，首席建筑师Enric ruiz-geli更装设多达300颗侦测器，连接触控式启动光源，透过感应人体活动来点亮灯光。此外，Media-TIC大楼还有一项设计重点，就是结合海水冷却的空调系统，善用相距不远的地中海，抽取摄氏10度的冰冷海水，再经由特殊的回路来冷却大楼内空气，同时将较高温的水循环导回海中；这么一来可减少大楼内逾七成的碳排放量，节能效益卓著，因此一落成即获得纽约时报等国际性媒体广泛报导，甚至得到全球建筑物节能最高荣誉LEED金奖殊荣，堪称世界最具指标性的智慧绿能建筑，这个奖项连我们的101大楼都在积极申办当中。

当京都议定书的减排协议即将于2012年届满，即将到来的哥本哈根高峰会势必要更加积极，无论是政府或民间团体，都亟需更广泛的新思维和新作法，让台湾能走向绿能国家的前驱。

建构线设计有限公司设计总监　　沈志忠

Introduction:

Horrible news caused by extremely bad climate has forced people to face up with the urgency of environmental protection, while this kind of trend allows various conceptions and actions of slowing down the pollution to extend from daily life to every space, architecture and environment, and also bring the new words like green wisdom.

Text:

As a professional designer, I have a very direct sense towards environmental protection. Surely, it's not difficult to carry out the spirit of environmental protection in the social unit of individual person or family, while as for big sized factory or business space where a lot of people are inside, since the combination of wisdom and low-carbon are the trend of future architecture, the topics of related environmental protection should be proceeded under the power of expertise and technology. Such as the electricity efficiency, power safety, light power+ artificial lightening, planting, water recycling, cold room efficiency, these are all the key points of big sized factory or business space. And the content about power can permit the owners to control the operation cost easily through the requirement management and power monitoring.

As energy-saving wisdom is becoming the mainstream of future market, every country in the world is striving to promote it, just like Media-TIC building located in Barcelona-Spain which is jointly created by local government and famous architect Enric ruiz-geli. This building is not only an innovational center, but also the sample of architectural energy-saving. Technical appearance is made of Teflon coating which is used also in Water Cube. This material can insulate ultraviolet efficiently and make good use of heat insulation. Furthermore, chief architect Enric ruiz-geli installed 300 detectors to connect touch typed switch to turn on the light which can be lightened through sensing human body. Besides, there's still a key point of design inside Media-TIC building, that is, combining the seawater cooling A/C system which adopts 10 ℃ icy water from not so far Mediterranean Sea to cool down the air inside the building through special loops, meanwhile, the hotter recycling water will be sent back to sea. Doing this can obviously reduce 70% carbon emissions of the building, which was reported widely by international medias like New York Times, and also granted the highest LEED golden award of global architecture energy-saving. It really plays the role of symbol of wisdom green building. By the way, this award is being chased by us for our 101 building.

The emission reduction agreement of city protocol will expire in 2012. The upcoming Copenhagen summit is bound to be involved more energy, no matter government or non-governmental organizations will need more and more new thoughts and actions to make Taiwan become the pioneer of green energy in the whole world.

Contents | 目录

006	New Office of Unliever Headquarter	联合利华总部新办公室
020	Engine Reception and Innovation Labs	Engine创新实验室
026	SKYPE – New Office in Stockholm	Skype斯德哥尔摩新总部
032	Prayer Rahs notary's office	Prayer Rahs公证办公室
038	BASF Headquarters BASF Headquarters	巴斯夫总部
044	Fashion Institute Taipei	时尚泉源之树—西园29服饰创作地
052	Remarkable Discipline of World	世之显学OF TBDC OFFICE
060	Royal Spirit Office	Royal Spirit 办公室
070	Bolon Office	波龙办公室
076	D.H.H Trade Company	D.H.H外贸公司
082	Student Loans Company	达灵顿学生贷款公司
088	Private Banking Centre of Alior Bank	Alior银行的私人银行中心
094	Bene Flagshipstore	BENE旗舰店
100	PKO Bank Polski Private Banking	波兰PKO银行的Polski私人银行分部
108	Red Bull Nederland B.V.	红牛荷兰公司
118	LEGO PMD Office	LEGO PMD办公室
124	ING Bank Slaski corporate department	ING银行斯拉斯科地区企业部
132	Zhen Shufen Design Consultance Company Hongkong Office	香港郑树芬设计事务所办公室
140	Dunmai	Dunmai办公室
146	A Bold New Office Environment	大胆的新办公环境
150	Forward, Office Extension	Forward公司办公室
156	Office BesturenraadBKO	BesturenraadBKO办公室
160	San Pablo Corporate Office	San Pablo办公室
166	Xi'an Maiyi Space Design Studio	西安麦一空间设计工作室

| 170 | A Dream Private Library in the Urban
都市丛林的梦想藏经阁 | 236 | 798 Art Factory
798艺术工厂 |
|---|---|---|---|
| 176 | WirtschaftsBlatt Newsroom in Vienna
维也纳WirtschaftsBlatt新闻中心 | 240 | Decoding DNA
解码DNA |
| 180 | Noble Bank
莱宝银行 | 250 | Office of ZhengMao Photoelectric Co.,Ltd, Shenzhen
深圳正茂光电办公室 |
| 186 | Open Finance
开放式金融 | 258 | Office at East Peace Road
和平东路办公室 |
| 190 | Le Cube Offce
乐立方办公空间 | 264 | Vinistyle Cosmeytic Office Building of Sumei Group
苏美集团vinistyle化妆品办公楼 |
| 196 | China Merchants Tower
招商中心 | 270 | Disonna's Factory Reconstruction and Indoor Design
迪桑娜厂房改造和室内设计 |
| 204 | Turkcell Maltepe Plaza Office
TURKCELL MALTEPE 广场办公室 | 274 | Kun'en Investment
坤恩投资 |
| 214 | Ymedia
Y媒介 | 280 | Santillana
桑提亚拉 |
| 220 | XYI OFFICE DESIGN
大隐于市的设计 | 286 | Cabel Industry
软件公司Cabel Industry |
| 224 | Youth Republic
青年王国 | 292 | Created Office
克里德办公室 |
| 230 | Splendid Shows for an Antiquated Space
大木和石loft空间 | 296 | Shanghai Oulin Office
上海欧林办公室 |

New Office of Unliever Headquarter
联合利华总部新办公室

设计单位 | Camenzind Evolution   项目地点 | 瑞士   摄影 | Peter Würmli

A day can be divided into 24 hours, one of every third of which is occupied by the works in office. Provided that the office's environment is too poor for people working there, not only has people's visual sense been affected, but also the working effectiveness and zealousness would be weakened.

This project is to create a second home by using warm colors and matching with large-sized irregular threads for the employees on the premise of keeping leisure and comfortable, then finally create a warm and harmonious office.

一天有24小时,其中三分之一的时间都是在办公室工作,如果办公室工作环境不佳,对工作的效率、工作热忱都会有影响。

本设计方案以休闲、舒适为前提,为员工缔造另一个家,以温馨的色调,配以大面积不规则纹路,营造出一个和谐、温暖的办公室。

Engine Reception and Innovation Labs
Engine创新实验室

设计师 | Jump Studios (Shaun Fernandes, Markus Nonn)　　项目地点 | 英国伦敦　　建筑面积 | 914 平方米

Two circular seating clusters offer seating space for up to eight people each. They have been kept simple in terms of colour and shape in order to harmonize with the multicoloured wall behind.

Labeled as the 'Innovation Labs,' these spaces are designed to offer alternatives to the more traditional meeting rooms on the upper floors. They have been conceived as rooms for workshops, brainstorming sessions, seminars and presentations and offer different ways of working, communicating and presenting to each other.

Innovation Lab One features two flexible clusters of low seating which can be rearranged to suit different numbers and sizes of groups. Together with four small meeting tables and a wealth of round stools in different colours this is an ultimately flexible space for working in smaller groups between two to six people. In addition there's a long workbench with high stools along one of the side walls for more individual work on computers or laptops.

The last of the three rooms, Innovation Lab Two, is yet another space. It consists of four round, red Waltzers which spin around. Jump Studios conceived this as a highly energetic area for workshops and brainstorming sessions. Four smaller groups can work individually and with a higher degree of privacy when the Waltzers are rotated facing the corners of the room. By spinning them 180 degrees to face each other the space can be transformed to facilitate open discussions and

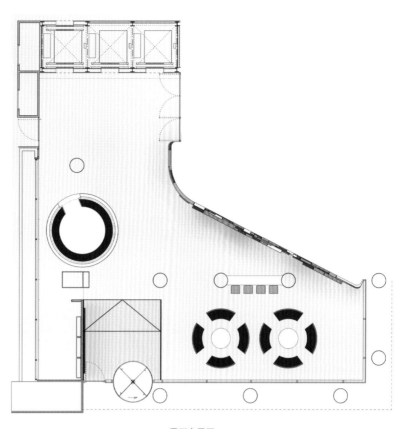

平面布置图一

presentations to the other three groups.

A connective element between all three rooms is a series of identical display screens facing the street facade. These elements play with notions of transparency and opaqueness, concealing and revealing. Made from translucent acrylic they filter the sunlight shining into the rooms during the day and reveal their inner grid like structure. At night the displays become glowing boxes when viewed from the street allowing peeks into the inside through round cut outs which act as windows, graphic displays, open shelves or LCD screens.

  两个圆形座位群各自可为八个人提供座位。它们的色彩和形状简单，以便和后面的多彩墙面保持协调。

  这个项目取名为"创新实验室"，这些空间被设计成传统会议室的备选方案。它们被视为是工作间、集思会议室、研讨会、展示会，以及提供不同的工作、交流和呈现方式的场所。

  创新实验室一的特点在于有两组灵活的矮座位，它们可以重新排列，供给各种人群使用。四个小型会议桌和大量颜色各异的圆形凳子组成了一个非常灵活的工作空间，供二至六人组成的小型工作团队使用。另外，沿一面边墙有一长长的工作台，带高凳子，给更多人用电脑或者笔记本电脑独立工作。

  三个房间的最后一个，创新实验室二，是另一个空间。它由四个圆形的、旋转着的红色Waltzers组成。Jump工作室把这看成是一个非常有活力的工作空间和集思会议室。四个较小的工作团队可以独立工作，当Waltzers朝着房间角落旋转时有着高度隐私。当它们以180°旋转的时候，可以轻松地对着另外三组进行开放式讨论和演讲。

  三个空间的连接元素是一系列朝着大街的相同的展示屏幕。这些元素在透明和不透明、隐藏和显示之间游移着，由半透明的丙烯酸制成。白天它们将进入房间的阳光过滤，显示内在的结构网格。晚上当人们从街上通过窗户、图示框、开放架子或者LCD显示屏望向内部时，这些展示物品变成了闪光的盒子。

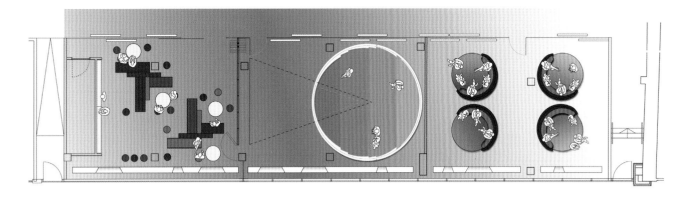

平面布置图二

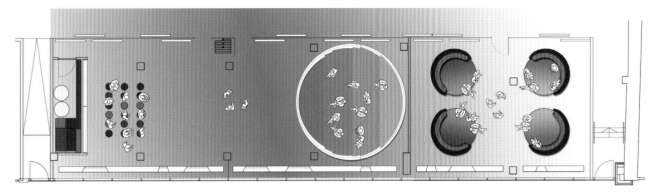

平面布置图三

SKYPE – New Office in Stockholm
Skype斯德哥尔摩新总部

设计师 | Peter Sahlin、Mette Larsson-Wedborn、Beata Denton、Therese Svalling and Erika Janunger 设计单位 | PS Arkitektur
项目地点 | 瑞典斯德哥尔摩 建筑面积 | 1680平方米

The project consists of audio and video studios, work space, meeting rooms and social areas for staff. The core idea behind Skype software application has generated the design concept for the interior of the new office; the idea being that Skype is a useful and wonderful application that allows chats, voice and video calls over the Internet on connected devices. A pattern has been derived from an idea of voids emerging between the abstracted interconnected nodes symbolizing the interconnected world. The scheme of this abstraction replicates itself in the flooring and in the design of the fixed interior. The idea of the bubbly furniture has evolved from the Skype logos. The Skype cloud known from the Skype logo, has literally been reinterpreted as a cloud-shaped lighting fixture, shining throughout the chill out space. The lighting fixture, truly one of its kind, is created by a cluster of lit up translucent globes of various sizes.

Created within a former brewery, a major effort has been made in order to accomplish high-end acoustics in the premises with efforts such as installing and designing soft wall absorbers. These efforts have been necessary for an office that predominately works with audio- and video development. This focus on audio- and video development is visible in the interior and expressed in the unique wallpapers with prints of cables, earphones and other devices linked to the audio-video technique.

The ebullient atmosphere and the vibrant colors are a direct translation from the graphics of the Skype brand. The contemporary interior generates not only an eye-pleasing environment but also an inspiring working place. Instead of the old fashioned offices with cubicles, this new interior creates a playful atmosphere that allows good, crazy and brilliant ideas to develop. An office should encourage a playful atmosphere and the Skype office truly does that.

　　本项目设计内容包括视频与音效播放室、办公区域、会议室与员工休闲区。Skype 总部内部设计的概念源自Skype软件应用的核心理念：通过一个集聊天、语音与视频通话功能于一体的因特网连接设备来展现其高效、完美的应用。而这一模式源自一个新兴的理念——抽象的、象征互联世界的互联节点。抽象概念的模式是将其本身复制在楼层的室内装饰上。气泡造型家具的概念演化自Skype的标识，被准确地解读为云形照明装置，照亮整个空间。照明装置中有一种类型，产生于一连串不同尺寸的、光亮的半透明球体。

　　总部所在建筑前身是一家酿酒厂。此案设计的一项重大努力是完成高端的声效设计，包括安装与设计软墙吸音板。这些工作都是为体现总部的音效、视频等工作发展所打下的必要基础。对于声效、视频发展在内部设计上得到体现，具体表现为墙纸上印有的独特的耳机、网线与其他音视连接技术的图像。

　　这里热情洋溢的氛围与振动式的颜色是Skype品牌图像的直接诠释。现代式的内部设计所催生的不止是养眼的环境，更是富有激发力的工作地点。这里不再给人老式办公室方方正正的感觉，新的内部环境是一个释放激情、才智的欢快场所。这里鼓励快乐，Skype做到了这点。

Prayer Rahs notary's office
Prayer Rahs公证办公室

设计单位 | Sue Architekten   建筑面积 | 390平方米   摄影 | HERTHA HURNAUS

Two notaries in Vienna are founding a new chancellery together and want to bring a fresh contemporary wind to their office.

Initial discussions were held, in which we also helped, in order to find the right building. A 390m² skyloft in a former telegraph office with a stretched thrilling space was selected: a space growing from a relatively narrow entrance area to a generous open loft space. The draft was honed in numerous meetings with the clients, the goal being an office that not only creates a communicative climate, but also enables concentrated and uninterrupted work.

The result is a layering of the office structure that maximises the potential of the location: at first a friendly entrance area to receive clients, followed by a general office and working area to provide an intimate and concentrated, while still relaxed, communicative atmosphere.

The focus in the central working area is on communication between staff. Stretched storage furniture separates the working area from the hall and communication zone. On one side there are generous cubicles and a meeting room, on the other side of the hall are se-ating niches, space for photocopiers and a common working area that is shaped by storage furniture. Small windows from the hall allow glimpses into the working areas. The other side of the hall leads to further individual offices.

The layering between the entrance area and the central working space is additionally supported by varying material and colours. Sand-coloured walls and floor tiling dominate the friendly recep-tion, the furniture used to define the rooms are a dark oak. White room furniture with textile patches, custom-arranged in each room, dominate the central area.

The generous impression of space provided by the skyloft is maintained, despite the visual and acoustic separation of the rooms from each other.

维也纳的两个公证处正合作建造一个新官署，旨在给他们的办公室带来新鲜的现代风。

我们参加了首次研讨会以找到合适的建筑大楼，最终挑选了前电信大楼里的一个390平方米带阁楼的延伸空间：从一个相对狭窄的入口区空间延伸出一个宽敞开放的阁楼空间。该方案是与客户多次协商后确立的，我们的目标是打造一个有着良好沟通氛围的办公室，同时，也能开展集中、不间断的工作。

最终，通过设计师合理的布局，多层次的办公室结构最大限度地发挥其位置潜能：入口区用来接待客户，随后就是办公室和工作区，紧凑集中而又拥有轻松的沟通氛围。

中央工作区是这个项目的重点，要求方便员工间的沟通。拉长的存储柜将工作区与大厅和交流区分开。一方面保证多间办公室和一间会议室的需求，同时大厅又可设置等候区、影印空间和公共办公区。从大厅的小窗口可以一瞥工作区的环境，而大厅的另一侧通往私人办公室。

入口区和中央工作区之间的隔层特别采用不同的材料和颜色来装饰。沙子色的墙壁和地板衬托着热情的接待台，房间家具由深色橡木制成。而每个房间带装饰织品的白色家具是特别定制的，占据整个中央区。

尽管每个空间都采用了隔音设计，视觉上彼此独立，但空间原有的整体形象有增无减。

BASF Headquarters
BASF Headquarters 巴斯夫总部

建筑设计 I SPACE、Juan Carlos Baumgartner、Gabriel Téllez Galindo　　设计团队 I Humberto Soto、Enrique Martínez、Elena Schneider
建筑面积 I 5,000平方米　　项目地点 I Insurgentes Sur Avenue, Mexico City　　摄影师 I BASF Courtesy

The BASF corporate building is situated in Insurgentes Avenue, one of the most famous avenue in Mexico City. The ground floor is of 1,315.0m² and each standard floor of approximately 942.0㎡. The shape has three differently-oriented squares brought together by a central core.

BASF's corporate office project consists of the ground floor and 6 levels, the second floor will be for leasing and the first floor will be developed in the future. The design program for the other floors includes spaces such as a lobby, open and private offices, meeting rooms, a dining room, rooms for informal or casual meetings, an auditorium, support areas and sites. A requirement of the client was to achieve LEED certification, so the design must had to incorporate maximum parameters of energy efficiency and sustainability.

On the ground floor there is a showroom for the display of the company's own products with the facilities to project images of the different products on screens.

Also located on this floor is the reception area which has a staple with a stone finish and the BASF logo on the back wall. With its waiting area and turnstiles with scanners giving access to users, it will service the entire building.

Restrooms are divided into two parts, one for the meeting rooms and the other for the dining area, executive dining room and

showroom. They will have everything to serve the needs of the customer with water-saving equipment which will help consumption. As this equipment has its own water treatment plant for used water, it will have a second use generating water for the air conditioning.

It has an office area and a space for medical services, a lactation room and an ATM.

The pedestrian entrance to the building has ramps allowing access to disabled persons in wheelchairs. The smoking area is in the same area, designed to have lower visual impact, so only the floor finish changes.

The floors second to the sixth could be called standard floors, whose three wings are comprised of open areas, except for the sixth floor where the presence of the chairman's office results in more private offices.

These floors have a closed printing room with extraction as well as a stationery storage cabinet, enclosed phone booths consisting of a seat and a telephone line for private telephone conversations between two or more people. These floors also have meeting rooms, mainly for 4 to 8 people, which will be equipped according to need and size with projectors and screens. There is also a janitor's room in the emergency stairwell, for storage and cleaning facilities.

A distinguishing feature of these floors is the informal meeting space, strategically located to serve coffee and water, so that people to gather casually and watch presentations run from a laptop set on a table.

Each informal meeting space is decorated with a different theme, in order to bring something special to each floor.

Low partitions between the workstations allow us to create a sense of spaciousness. Unlike the ground floor where the ceilings are either smooth or in a grid, these floors have shim plates and a center soffit grid, creating more movement and depth as you can see the slab behind.

On the sixth floor, the directors' area has a smooth drywall ceiling with recessed linear luminaries with louvers; these fixtures will have different intensities and scenarios.

The elevator lobby will have a smooth ceiling and a light box along its length to round off each wing of the building.

The lighting designer's proposal for the open areas is to have randomly suspended luminaires at two heights, adding a touch of versatility as this form of lighting provides for greater uniformity of lighting throughout the space.

The air conditioning works with Fan & Coils which enables us to make as much use as possible of the ceiling space, and thereby to give a greater sense of space to each location.

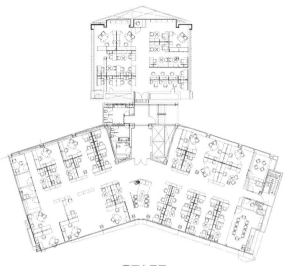

平面布置图

巴斯夫公司大楼坐落在墨西哥城著名街道——Insurgentes大道。占地面积为1315平方米，每层楼面面积约942平方米。三个不同朝向广场一起汇聚成大楼核心。

整个办公大楼有7层，包括一个地下层，二楼用来出租，一楼为未来的开发设置。其他楼层的设计方案包括大厅、开放式办公室、私人办公室、会议室、餐厅、非正式会议室、礼堂、支持场域。客户要求达到LEED认证，因此，设计必须实现能源有效利用和可持续发展。

地下层的陈列室用来展示公司产品，包括产品的投影展示。同样位于该层的接待区，其接待台以石材饰面，后墙上是巴斯夫标志。等候区旋转门旁的扫描机供整个大楼用户使用。休息区分为两个部分，一部分是会客室，另一部分是用餐区、行政餐厅和陈列室，能够满足于顾客一切需求，带污水处理器的节水设备回收利用空调排水。另外，还有办公区、医疗服务区、哺乳室和ATM机。

大楼人行道入口处的坡道方便坐轮椅的残疾人进入，为了产生最小视觉冲击力，同一区域的吸烟区只是地面颜色不同。

第二层至第六层是标准楼面，除六楼外，其他皆是三面敞开式设计，六楼的设计是为了满足董事长办公室更多的私人空间。这些楼层有一个带外置文件储存柜的封闭印刷室，带座位的封闭电话亭。同样，这些楼层设有会议室，可容纳4~8人，据需求和空间大小配备投影仪和屏幕。紧急通道口是门卫室，带储存空间和清洗设施。

　　这些楼层的一个显著特色是非正式会议空间，有效利用空间配置茶水间，方便人们随意聚集和观看幻灯片放映。同时，为突出每个楼层的与众不同，每一个非正式会议空间都饰以不同的主题。

　　工作站间的低分区带来一种宽敞感。不同于地下层光滑或网格天花板，这些楼层都有垫板和一个中心拱腹电网，使空间更富动感和纵深感。

　　六楼董事区采用的是光滑的石膏板吊顶和百叶槽线性灯具，灯具有着不同的照射强度和光感。电梯内配置光滑的天花和灯箱。

　　开放区随机打开两种不同高度的灯具，这种形式的照明让整个空间的统一照明多功能化。

　　空调与风扇、线圈搭配使用，尽可能多地使用天花板空间，从而提供更大的空间感。

Fashion Institute Taipei
时尚泉源之树——西园29服饰创作地

设计师 | 沈志忠、李怡霖　设计单位 | 建构线设计有限公司　项目地点 | 中国台北
建筑面积 | 1F=291.62平方米、2F=271.96平方米、3F=303.62平方米
主要材料 | EPOXY、喷漆铁件、防火皮革、人造石、清玻璃、PVC地板、柚木木皮染色

Clothing design gives consideration to the shape and skin of the human body, and uses materials with organic compounds. Therefore, we set one of nature's life forms –"Tree" as this project's theme. Wanhua Fashion Culture Museum will transform into the most important tree in Wanhua, and cultivate international fashion design talents in Taiwan.

The design concept of the space was to use an organic form to symbolize diverse development and creative energy of fashion design. When you enter the reception space on the first floor, it is like entering into a tree; the spatial concept is based on the structure of a tree, and spatial functions are created using growth rings. For the vertical image, the concept of "vascular bundle" of plants is applied; vascular bundles transfer nutrients and energy. Thus, we define vascular bundles as lights, emitting the energy of light into the space.

The studio on the second floor starts out as a hive in the tree, indicating that its fashion designs must be careful and thorough like a bee hive. We built a glass hive in the space of the large tree, where crystal seeds are nurtured –rising stars of fashion design. This is the

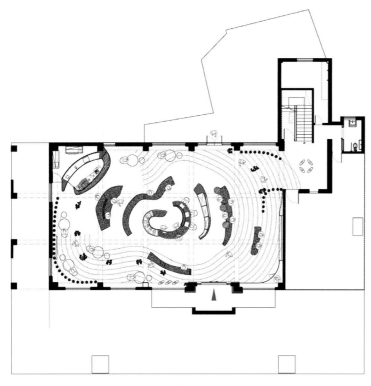
平面布置图一

平面布置图二

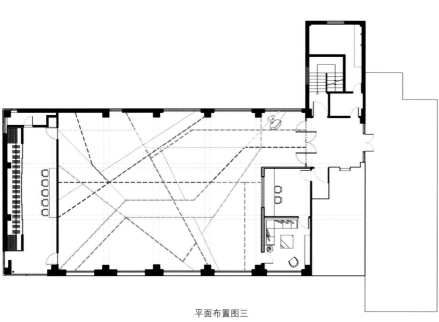
平面布置图三

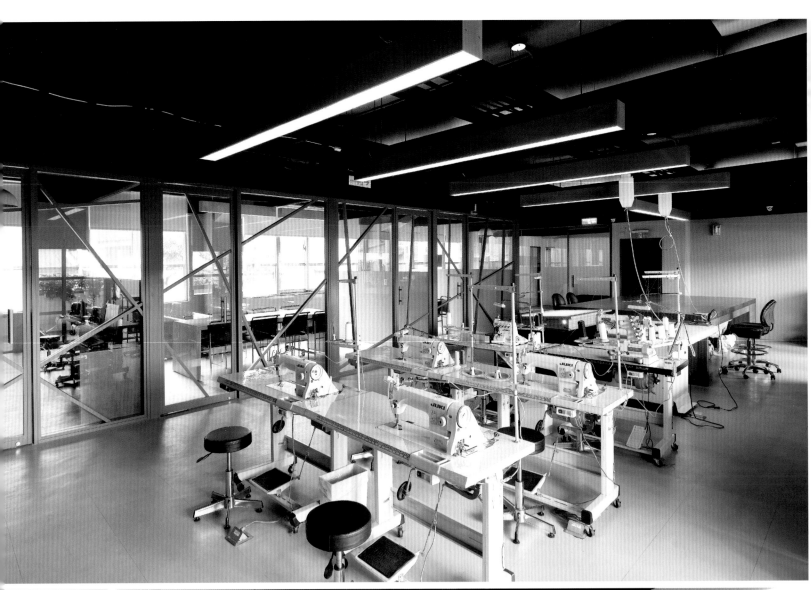

starting point where designers find the most suitable angle and elements of the culture museum, and let them penetrate the space, so that the creativity and influence of the fashion culture museum naturally flows within the space.

The third floor is a multifunctional activity space designed using the streets of Wanhua District as a blueprint, turning streets into tracks on the ceiling. The design is based on the concept of multifunctional and flexible utilization, and allows space to serve diverse audiovisual and exhibition functions, turing it into a performance stage with maximum effectiveness.

　　服装设计是以人体的形体与肌肤为考量，所运用的材料也都蕴含着有机体，因此设计师设定大自然的生命体——"树"为此案的设计主轴。万华服饰文化会馆将化身为万华最重要的一棵树，扮演着培育台湾国际服装设计人才的角色。

　　本案空间的设计概念以有机的形态出现，象征服装设计的多元发展与未来蓄势待发的创意能量。进入一楼接待厅空间犹如进到树的内部，空间概念取自于树体的结构，借由年轮的轮廓组构空间的形体，赋予空间机能；在垂直意象上，采用植物体的"维管束"概念，维管束具备养分与能量传输的功用，在此我们将维管束定义为发光体，将光的能量传送到空间中。

　　二楼的工作室区域，以"树中巢"的象征意涵起头，意指服装设计须有蜂巢式缜密的心思。设计师在大树的空间中建起一格格的玻璃巢房，巢房里孕育着结晶种子——服装设计新秀，设计师以此为出发点，找出文化馆空间里最适合的角度与因子，让它们贯穿于垂直水平的空间中，服饰文化馆创造力与影响力自然散发于其中。

　　三楼为多功能活动展示空间，以万华地区街廊作为构想，将街道转换为展板的挂轨反射到天花板上，多功能弹性运用的概念设计，让空间可以因实际应用多元变化视听与展示机能，使展演舞台达到最大使用效益。

Remarkable Discipline of World
世之显学 OF TBDC OFFICE

设计师 | 黄鹏霖Janus Huang、黄怀德Roy Huang　　设计单位 | 台北基础设计中心 TBDC
项目地点 | 中国台北　　建筑面积 | 1F 182平方米　B1F 90平方米　　主要材料 | 红橡木皮染色、集层夹板、结晶钢烤门板、不锈钢扁铁板、黑烤玻璃、黑镜、灰镜、PS板流明天花板、黑闪电大理石、仿砖面PVC地砖、氟碳烤漆铁件、4CM透明亚克力板　　摄影 | 王基守

The most remarkable discipline of the current world is design.

The design of this case started in 2009 and was completed bit by bit in 2012, during which a lot of adjustments were made. This is because, compared with the past, design has become famous modern doctrine, which correlates closely with real life. Owing to the downgrade of people's action competence caused by the convenient network information, the frequency of real contact with outside world slows down. In order to promote human feeling towards things and then stir and accumulate some experience from this feeling, TBDC hopes to improve the vitality of the designers and change the traditional mode and thinking habits of design in the process of planning hardware. This idea dominates the main pattern and spirit of spatial design.

Water template acting as the main measuring tool, combined with the vertical solid wood and horizontal imported brick material, the outdoor creates the atmosphere of spaciousness at the passageway and guides the sigh line to the indoor. The spacious French window separates the rest area from the large interact area. The introduction of natural light which enrich the space, the application of automatic curtain working as sunshade and projection, double automatic door, independent guests porch and the widely use of mirror surface as

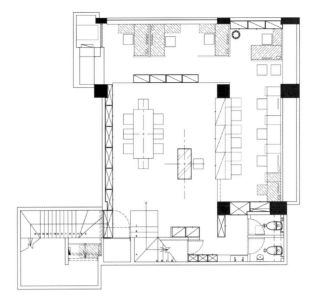

平面布置图一

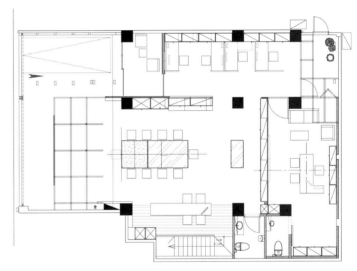

平面布置图二

the leading material which creates three-dimensional reflection effect make the space free from constriction. The outdoor day lighting landscape, which connects the general management area and the CEO workroom, satisfies the lighting of the two districts and the ventilation of fresh air. The covert projection equipment in the CEO workroom appeals the owner's need of privacy. The four-layer overlapping wooden bookcase working as the extension of the main interact area connects the multi-media interact bar area and ground floor.

TBDC encourages designers to get off their seats to ask the users and experts in other fields to take part in the design. During this process of spatial moving and interacting, together with music and coffee fragrance which set the tone for the space, future design discipline will take shape.

　　世之显学，设计也。

　　本案的设计开始于2009年，直至2012年才陆续完成，过程中不断地修正调整。因为发现，TBDC相较过去，设计已然是现代显学，同时与现实生活联系密切，TBDC认为设计师的思维需要改变，网络信息的发达，虽然便捷，却也造成人的行动能力下降，从而与外界实质接触的频率趋缓，提升对人事物的感觉。TBDC希望从硬件的规划，提升设计工作者的活动力，改变设计工作的传统模式与思考习惯，这个思维主导了空间设计的主要格局与精神。

　　TBDC将个人工作区域精简化，同时将互动区域最大化，并设置在主要空间的中心，一楼的互动区域设置了三区，地下室互动区分成两区，因不同的互动需要交替使用，作为设计会议、多媒体互动分享、材料与设备选用以及翻阅图书数据等。除了可以不定期邀请各领域专家举办座谈，同时希望配合每个设计元素不同的流程，在这些空间的业主、设计团队及供货商等，都能近距离接触，直接作视觉、听觉、味觉、嗅觉、触觉等的传达与联系。

　　户外以清水模板作为主要量体，结合垂直向实木与水平向的进口砖料，营造出入口的宽敞与引导视觉进入室内；以大面落地窗作为入口休憩与大型互动区的区隔，引入自然光线滋润空间表情；自动化卷帘作为遮阳与投影使用，双自动门独立迎宾玄关；使用大量镜面作为主要材料，三度的反射效果使空间没有压迫感；总管理工作区与CEO工作室由户外采光景观作连贯，同时满足两个空间的采光与新鲜空气的流通，CEO工作室内设置隐蔽的投影设备，满足业主隐私需要。四层交迭错落的木质大书柜，作为主要互动区域的延伸，连接多媒体互动区域。

　　TBDC鼓励设计工作者离开自己的座位，引导使用者与其他专业领域专家参与，在这个空间移动与互动的过程中，伴随着音乐与咖啡香，为这个空间定性，激荡出未来的设计显学。

Royal Spirit Office
Royal Spirit 办公室

设计单位 | 蔡明治设计有限公司（ALEXCHOI design & Partners） 项目地点 | 中国香港 建筑面积 | 4004平方米
摄影 | Alex Choi、Kit Siu

The industrial re-revolution is happened in textile office located at a 70's Factory Building which carries a long history. Back to the 70's, Toys, Textiles, Watches and Jewelleries, are the 4 king industries that had brought Hong Kong an economic summit. But time flies, the heroes have been forgotten like they have never existed.

Born in 70's, designer's mission is to bring the classic back and raise the industrial re-revolution. This textile company needed a repackage service to hark back people's memory and meet the trend. Like a quantum leap: 2 extreme elements meet, it causes a big bang.

Primary material is constantly used in the space, like iron, steel, glass, brick... these materials maybe common to the world nowadays, and people also take them for granted. But the fact is, these materials are used to be great inventions that had stunned the world, and they are always playing their part by improving peoples living. These harmonious materials create a minimal ambience by only using black, white and grey as the theme colour of the space, keeping the area true and humble with no extra accessories when the natural materials are the best accessories, and the use of dramatic lighting shines out the true beauty of the materials.

本项目是为服装企业而设计的。该场所位于一幢历史悠久的20世纪70年代厂房。玩具、手表、服装、珠宝，曾几何时，促成了香港的经济腾飞。可时光飞逝，现如今这些"英雄"们早已被遗忘，就像从来没有存在过一样。

生于70年代的设计师们的任务就是让经典回归，重启又一次工业革命。服装企业急需重新包装其服务，以追回人们淡忘了的记忆，也能满足潮流的需要。就如两个相斥的元素碰在一起，引发大爆炸。

主要材料如铁、钢、玻璃、砖，被不断运用在空间当中，这些材料或许普通常见，易于获得。但事实上，它们曾是影响人类世界进程的发明，对于人类生活的改进它们贡献繁多。这些和谐材料运用白色、灰色作为主色调将空间变得真实、平和，没有过多的装饰。材料本身的自然品质就是最好的装饰。此外，灯光也将材料的真实美感完全尽现。

Bolon Office
波龙办公室

设计师 | 黄书恒，玄武设计群　设计单位 | 玄武设计　项目地点 | 中国台北　建筑面积 | 245平方米
主要材料 | 波龙毯、壁纸、玻璃隔断　摄影 | 王基守

Bolon Art, being famous for its unique weaving techniques, gives the clients an experience of phantom and reality with the colors and dance of lines through the complicating textures. Such a deep intrinsic of the enterprise, as is both a phantom and a reality, becomes the start point for the designer's plan.

The designer made use of white as the main color for the wall and the ceiling because its bright and swift could be good for visitors. The white is the origin of every colors here and the designer had considered it as the foundation for the ideas as well as use it as the exclusive color in the office. At the mean time, the background in white has successfully optimized the products' arranging effect either. All the weaving samples put in the shelf seem to be the perfect works of art. The weaving carpet in ground color has floored everywhere to match the vertical carpet floored alongside the wall. Therefore, the carpets can be in conform with the sprit of enterprise. Nevertheless, the carpets can also supply the visitors a feeling of secures. In the main office area, the designer chose blue glass to separate the space. In one hand, the blue, altogether with the pure white of the hall and the iron gray of the main office, maintains the relaxing feeling of the conversation with the clients as well as the rigorous working attitude of the staff's. In another hand, the swift blue enlarges the sight and create a more free atmosphere in the space.

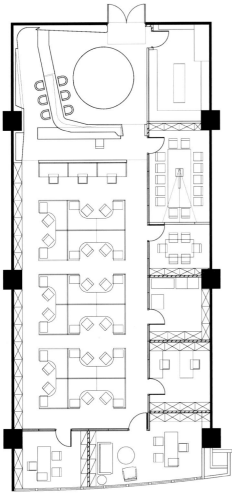

平面布置图

In this project, not only had the designer played a excellent planer for the color using, but also made a lot of efforts for the lining art. In the conversation area, the smooth arc lines enhance the vivid feeling of the ceiling. The arc lines go through the wall till the adjoining corner and turn straightly down in order to make the function of the wall subtly. Therefore, the sample shelves, boards and the desktops metled into one. It can be seen that the designer showed an active imagination not by the excessive use of the technique but by the sample and clear lines. Making the visible lines into invisible ones to achieve the endless expanding effect, creating the unreal interactive relations with the unstopping and turing lines, the designer brings the deep intrinsics for the people in the space.

"波龙艺术"以特殊的编织技术闻名业界，织毯上的绣线纵横交错，利用独家技术呈现繁复纹理而不显累赘，让使用者藉由色彩与线条的轻舞，逡巡于真实与虚构之间，这种若有似无、如真似幻的企业内蕴，便成为设计者规划此空间的出发点。

设计师采用白色作为墙面与天花色泽，以明亮与轻盈感浸润访客的感官，"白色"是每种颜色的起源，设计师使之成为办公室的统一色泽，隐喻着工作者萌发创意的基底，同时白色背景也让产品的摆置效果倍增，多采多姿的织锦样品整齐置于架上，俨然成为一座独立艺术品。以大地色调的织毯铺满洽谈空间，与墙面摆置的横向地毯互相呼应，不仅符合企业精神，同时也提供给访客"脚踏实地"的实在感受。走近主办公区，设计师选用蓝色玻璃分隔内外空间，与门厅的大片纯白，亦接壤主要办公区的铁灰，维持与访客商谈的轻松感，亦无损工作者应有的严谨气质；另一方面，轻透的蓝色也提升视觉广度，让空间动线默默逐渐延伸，营造更自在、从容的空间氛围。

于本案中，设计者不仅扮演着高妙的色彩策略师，也投注相当心神于"线条技艺"之中。走进洽谈空间，圆润的弧线提升天花板的活泼感，一路延展至墙面交接处便笔直往下，藉由持续起伏，灵动地勾勒出每面墙的功用——样品架、摆置画作、招牌、乃至与立体桌面巧妙化为一体，可以看出，设计师无意卖弄过多技巧，反藉由简单干净的线条，呈现深沈而活跃的设计想象，利用起伏、凸起和隐藏等各种视觉魔术，让有形的线条化为无形，达到无尽延伸的效果，不停盘旋、充满转折的线条，演绎着虚实的互动关系，让置身其中的人们游走于有无之境，享受设计者带来的丰富内蕴。

D.H.H Trade Company
D.H.H外贸公司

设计师丨竺李佳　设计单位丨宁波莱卡空间展示设计工作室　建筑面积丨1200平方米
主要材料丨亚克力板、雪弗板、不锈钢、彩色玻璃、抛光砖

Design note: As a case about office, designer always follows the principal of combining practical usage, functional requirement and humane management, which connects office requirement and work flow chart, and divides functional areas in a scientific and reasonable way, considering the communication between employees and leader, among functional areas, working and leisure. Material is simple and grand. Endurable and environmental friendly modern material is used in lightening to meet the requirement of office. After designing delicately, the space is clean, neat and beautiful, showing the image and modern feelings of enterprises.

The plane arrangement uses elevator hall to divide two big functional areas of working and entertainment, and organize the size of working area according to human body engineering and office furniture. The whole sections cannot only meet the requirements from different department and arrangement of human resource, but also combine every department to better the working efficiency. Working area is next to the elevator and toned with white. The reception hall is without too many bright colors and redundant ornaments. White paint and marbles decorate the reception desk, showing the designer's modern simplicity design theory. The whole design emphasizes practical, modern, environment friendly and humanity to create nobility, elegance and sobriety. Since the corridor is a little long,

glass walls are adopted and decorated with cubical sculpture English logos of company, matching by cold colored light to highlight its conceptual fashion. Every office uses characterized Arabian numbers to decorate its door. The office is clean and bright, with reasonable arrangement of light, at the same time, using the whole lightening together with individual lightening to form smart, bright and comfortable interior space. The employers' desks are short partitioned furniture which can connect numerous desks to a team and tactfully arrange in straight lines or slash lines, simple but grand. Besides, designer also designs coffee bar, KTV, cards room and gym, which can play the role of relaxing and entertaining place for employees after busy work.

　　作为一个办公空间专业设计方案，设计师在平面规划中始终遵循实用、功能性和人性化管理充分结合的原则，在设计中，既结合办公需求和工作流程，科学合理地划分功能区域，也考虑员工与领导之间、功能区域之间、工作与休闲之间的相互交流。材料运用简洁大方、耐磨环保的现代材料，在照明采光上使用全局照明以满足办公需要。经过精心设计，在满足各种办公需求的同时，空间简洁、大方、美观，充分体现企业的形象与现代感。

　　平面功能分布以电梯厅为界线，划分出工作和员工俱乐部两大功能区域，并按照人体工程学和办公家具尺寸对工作领域进行功能区域的大小划分。整个分区既能满足各部门的功能要求，对人员进行分流，又使各部门有机结合，合理利用空间，提高工作效率。走出电梯厅就步入工作领域，空间以白色为主色调。接待厅没有花哨的色彩和多余摆设，白色亚光油漆和雅士白大理石贴面装饰前台结合白色墙面，体现现代简约的设计理念。整个设计重点彰显实用、现代、环保和人性化，营造空间的高贵、典雅、庄重。由于走道较长，设计上采用玻璃幕墙，并装饰具有公司品牌特色的立体雕塑英文字，配以冷色调灯光，凸显走道时尚概念化。每个办公室门以个性化的阿拉伯数字作为装饰，办公空间色调干净明亮，灯光布置合理，在合理运用天光的同时采用整体照明与个别空间照明相结合，形成快捷、明亮、舒适的室内空间。员工办公桌选用矮隔断式家具，它可将数件办公桌以隔断方式相连形成一个小组，布局时将这些小组以直排或斜排的方法来巧妙组合，简单又大方。设计师还为员工俱乐部设计了咖啡吧、KTV、棋牌室和健身房，员工在忙碌的工作之余拥有了一个放松心情的活动场所。

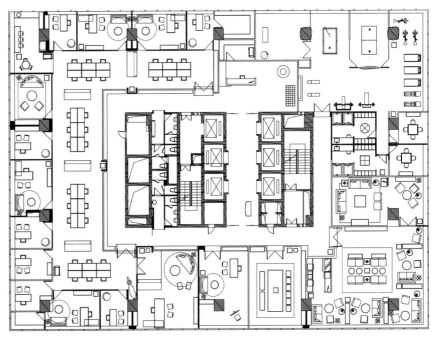

平面布置图

Student Loans Company
达灵顿学生贷款公司

设计单位 | Graven Images   项目地点 | 英国达灵顿   摄影 | Renzo Mazollini

We were commissioned in 2007 as part of a 3-year framework to design the interior of the new Student Loans Company contact centre in Darlington. The building was previously the Patons and Baldwin thread factory and this project forms part of the regeneration of Lingfield point. Student Loans Company wanted a building that would change the perception of their organisation; an aspirational new facility which would represent a new approach to their customer service, accommodating over 600 new and existing staff, and to assist in attracting and retaining staff among a competitive and youthful market place while retaining the heritage and historical elements of the area.

We were lucky that we had the 'Memphis' building with its great history and character as a starting point, with a great natural light quality and high ceilings. The central circulation zone referred to as the 'Street' comprises the reception and welcome area, meeting and training rooms, learning and coaching zones, meetings booths, complete with cafe and dining areas.

The mezzanine space above this created an additional staff breakout and dining space.

The building has an industrial feel and is completely open plan. We tried to introduce innovative materials to the space and to introduce colour and texture not usually adopted within a corporate environment – this is not a typical office environment.

我们于2007年承接了这项工作。这项工作旨在对"达灵顿学生贷款公司"业务中心内部设计进行为期三年的框架工作的组成部分。这栋建筑的前身是帕通斯巴德文制线厂,这个设计则要将其改造成振兴灵非德区域业务的一个重要部分。

学生贷款公司希望通过本建筑改变外界对其公司的感知:一个充满灵感的崭新建筑,代表着对客户服务的新途径,能够容纳600名员工的办公空间,能帮助公司吸引和留住这个年轻、充满竞争活力的市场,也保留住这个地域的历史和珍贵遗产元素。

我们很幸运地拥有"孟菲斯"建筑和它的伟大历史与特征作为始点,这里面还有完美的自然光质和极高的天花板。中心的圆圈区域代表着"街",包含了接待和欢迎区、会议、训练室、学习、教练区、会面屋,还配备了咖啡厅和餐厅。

上层的夹层区提供给员工额外的休息和用餐空间。

整栋建筑充满了工业动感,完全开放。我们试图将新型的材料运用在空间里,将不常用于企业环境的颜色和材质用在这一项目上来,因为这并不是一处普通的办公环境。

Private Banking Centre of Alior Bank
Alior银行的私人银行中心

设计师 | Robert Majkut　设计单位 | Robert Majkut Design　项目地点 | 波兰　摄影 | Olo Studio

Alior Bank is one of the most dynamic brands of financial institutions in Poland, constantly expanding the offer and gaining more clients. Entering new area – the sector of private banking - required creation of an interior design that would respond to the highest level of service for the most demanding customer.

The main assumption of the project concerned the development of the very characteristic and strong bank's identification based on three colors: crimson, yellow and white, and is styled as an old drawing – the face of an angel. This element is so characteristic that it became the starting point for further graphic works and studies, which are included in the design.

The interior design concept of Private Banking branches had to be supplemented. It also required designing the elements, which would differentiate it from the retail branches. That's why black instead of crimson was incorporated into the sign. It creates a nice composition with other newly selected supplementary colors such as the shades of copper, grey and graphite.

The graphic linearity in the design appears not just in the rescaled logo dominating the reception zone, but is also included as the characteristic pattern on the carpet, multiplying the effect of

perspective in space and referring to the effects observed in the sign.

Another aspect of the design transplanted from the visual identification concerns the shape consisting of two arranged squares. As a starting point they became the leading theme of all forms and shapes of the interior reminiscent on one side of the paintings of Piet Mondrian and on the other side of the already classic Bauhaus school.

The rhythm of lines and geometry of the appearing and overlaying squares create the internal divisions, starting from the main space plan and finishing on the furniture detail of the conference table top. The linearity and rhythm of the squares became the pretext not just for the development of the forms but also for the purely decorative effects.

Alior银行是波兰金融机构中最具活力的品牌之一,它一直在不断扩大市场份额及获取更多的客户。为了进入新领域,该私人银行业务部需要重新打造室内设计为最苛刻的客户提供最好的服务。

在深红、黄、白三色的基础上,该项目主要关注其特色发展和身份标志,其标志选取古老的画作——天使的面孔。此元素极具特点,成为未来设计中选择美术作品的出发点。

私人银行分支机构的室内设计也需要加以补充,要求设计元素区别于其他零售分行,这就是为什么标志的颜色由深红色换成黑色。黑色与其他补充色,如古铜色、灰色和石墨色等,构成一个有机整体。

图形线性设计不仅体现在接待区重新规划的标志上,也体现在地毯的特征模式上。这样一方面增加空间视觉效果,也提升了标志的观赏性。

视觉识别设计的另一焦点在于两个正方形组成的外形设计。作为设计出发点,它们成为室内形式和形状的主题,一面是蒙德里安的画作,一面是经典的包豪斯学校。

线条的节奏感、几何外形及叠加的正方形对室内空间进行了划分,从主要空间规划直到会议桌的顶部细节。另外,正方形的线性感和节奏感为形式的发展和装饰效果奠定了基础。

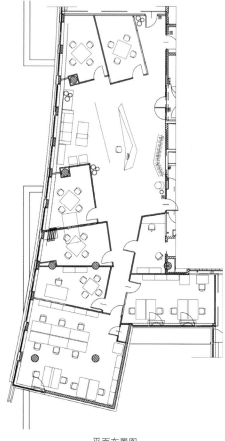

平面布置图

Bene Flagshipstore
BENE旗舰店

设计师 I SOLID architecture ZT GmbH   项目地点 I 奥地利维也纳   建筑面积 I 950平方米

With a floor area of 1000m², the showroom is located at ground floor level and forms a band that runs along the facade.

The intention was to offer visitors an interestingly varied tour through the world of Bene and at the same time to create a place with its own identity and an unmistakeable character.

Design of the showroom

The walls, floor and ceiling are the surfaces that determine the character of the space: for each of them a single material was used throughout.

Wall cladding – the "backbone"

The entire rear wall of the showroom is clad with panels covered in fabric and forms its spatial backbone. In the area that is also used for events the panels are backed with an additional sound-damping acoustic inlay.

Ceiling

The acoustic ceiling to the showroom is made of blown glass granulate. All the inserted services are flush with the ceiling.

Flooring – terrazzo

The entire showroom, including the service spaces, has a light-coloured terrazzo floor. It forms a continuous surface on which

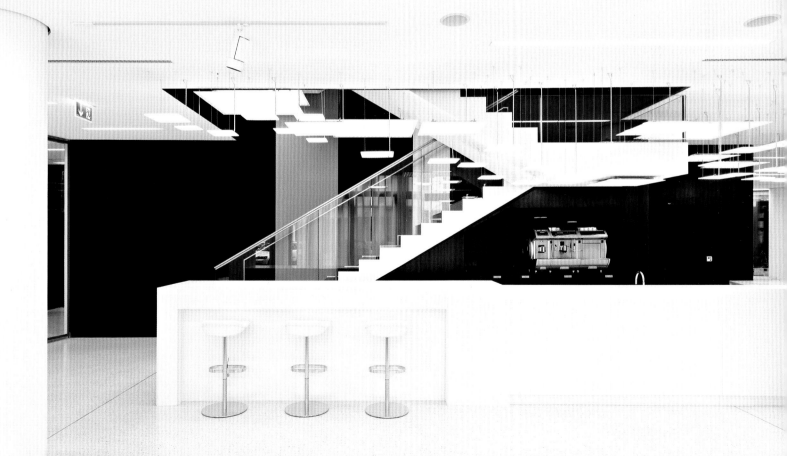

individual functions and presentations can be individually emphasized by means of carpets.

Colour scheme

In selecting the materials for the floor and walls it was important that these surfaces should provide a harmonious background for furniture with different surface finishes and colours. At the same time the showroom was to convey a warm and friendly underlying mood.

Consequently it was decided not to use monochrome surfaces in the design of the showrooms but materials made up of several colours that also have a high haptic quality. The large area of light-coloured terrazzo has a white, grey and brown natural stone "grain". The dark brown fabric used to cover the "backbone" also follows the same concept; the material is woven from brown and black threads.

Internal staircase

The core of the Bene premises is a centrally positioned internal staircase that connects all the floors used by Bene. This stair forms the central communication area to which certain gastronomic functions are attached: the bar with the "coffice" on the ground floor, and the tea-kitchens and the areas for informal communication on the two upper floors.

The internal staircase follows the design concept of the ground floor showroom. The staircase walls are clad with panels covered with brown fabric. In addition a "cloud" of square LED lights extends from the ground floor to the second floor.

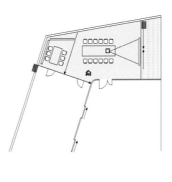
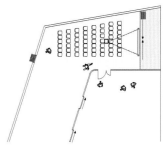

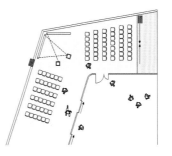
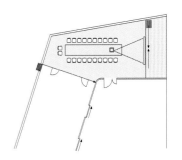

平面布置图

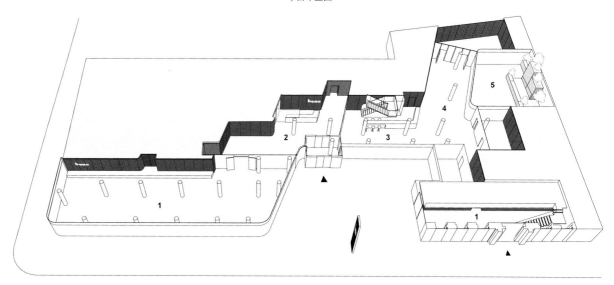

鸟瞰图

展厅设在地面层,沿着朝向街道的一面延伸,其建筑面积为1000平方米。

本项目设计意图是通过BENE的世界,为游客提供趣味盎然之旅。同时,创建一个完全符合企业形象的地方。

展厅设计
墙壁、地板和天花板为整个空间外观定性,因为这三者均使用单一的材料。

墙面——"脊梁"
展厅的整个后墙包布覆盖板,并形成空间的主干。镶嵌于墙壁上的隔音板也采用相同的方式。

天花板
展厅的天花板由具有吸声功能的吹制玻璃颗粒构成。所有其他的装饰需和天花板融合。

水磨石地面
整个展厅,包括服务场所,都有一个浅色的水磨石地面。这样整个地面就连在了一体,我们可以通过铺设不同的地毯来进行功能分区。

颜色方案
在选择地板和墙壁材料的时候,很重要的一点是,设计应该围绕创建一个和谐的家具氛围来开展,因为不同的家具,其表面光洁度和颜色也不同。同时,展厅需要传达热情友好的情绪。

因此,设计公司决定在展厅的设计上不使用单色表面,而采用具有高触觉感的、颜色各异的材料。大面积的浅色水磨石地面采用白色、灰色和棕色的天然石材颗粒。主干部分包布着深棕色的覆盖板,这也遵循相同的概念。这些材料通过棕色和黑色的线来进行编织。

内部楼梯
BENE设计核心的前提是用一个中心定位的内部楼梯来连接所有BENE的楼层。这个楼梯可以通到具备美食功能的区域,如地面层的"coffice"酒吧,茶房和两个楼上的非正式区域。

内部楼梯遵循一楼展厅的设计理念。楼梯墙壁被棕色板覆盖。此外,方形的LED灯群从地面层一直绵延到二楼。

PKO Bank Polski Private Banking
波兰PKO银行的Polski私人银行分部

设计师 | Robert Majkut   设计单位 | Robert Majkut Design studio   项目地点 | 波兰   建筑面积 | 483平方米

Robert Majkut Design studio, in cooperation with PKO Bank Polski, developed a project for private banking, offered to a selected group of the Bank's clients.

The starting point for this project was Bank's corporate identity developed by White Cat Studio, above all the modernized logo of PKO Bank Polski in its elegant color version for Private Banking sector - black, white and gold, which created a set of basic colors for the interior. Other graphical elements were also inspiring – in particular one decorative motif consisting of a grid of elegant, sinusoidal lines, consistently applied in the graphic design for the PKO Bank Polski Private Banking.

The theme of delicate grid was treated in a very innovative way – a bi-dimensional pattern was completely transformed by introducing an additional dimension: it was made spatial by being projected on the tri-dimensional model of the interior. To achieve this effect special software for parametric design was used, which allowed for the creation of a complex and ordered structure formed as a transformation of the subtle grid of lines converging in one abstract point. This complicated geometry setting has become a model to be filled with interior design solutions.

Corporate colors of PKO Bank Polski - black, white and gold - have been applied with full consequence, further complemented by shades of gray. The designer has taken an specific way to play with the tones of gold which are set in the interior in a kind of drama. Gold balances strong, graphic contrast between black and white and has unique decorative qualities and rich symbolism – as a warm and noble color referring to such values as prestige, stability and prosperity.

Identificational colors of PKO Bank Polski Private Banking brand appear in the interior in varying proportions and intensity. The customer service area, because of its specific character, is dominated by an elegant combination of black and gold, merely broken with white, but in the back-office area the leading color is white. The walls of meeting rooms are dominated with the color of warm gold, which, after the black reception hall and corridor, symbolizes getting into the essence and heart of the place. The black hides its real "inner" richness of gold, not dazzling, but slowly revealing its presence.

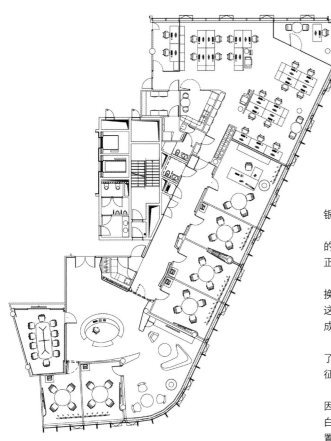

平面布置图

罗伯特·马伊库特设计室与波兰银行合作，开发了这个私人银行项目。本项目专为特定的银行客户而设。

项目的启动点是白猫工作室开发的银行企业身份。波兰银行私人银行部的现代化标识以尊贵的颜色——黑色、白色、金色作为内饰的基色；其他的图形元素也同样启迪心扉，尤其是优美的正弦曲线网格装饰理念，贯穿于整个项目的图形设计。

精致的网格主题被当作是一项创新——二维模式受被加入的一条额外的维度影响而被完全转换：它因内部的三维模式构建而呈现立体化。为达到效果，设计上使用了专门的参数设计软件。这个软件可以将在一个抽象点上浮起的微妙线条转换成复杂而有序的结构。这个复杂的几何设计成为了内饰方案的一种模式。

波兰银行的企业颜色是黑色、白色与金色，辅以灰色背景。设计师以特殊方式将金色调做了调节。金色将黑色与白色搭配所产生的强烈图像对比做了平衡，并产生了独特的装饰质量和象征——温暖、尊贵的颜色象征了威严、坚定和成功。

波兰银行私人银行部门品牌的身份颜色在内饰上呈现了不同的比率和集约度。顾客服务区因其特殊性，装饰以优雅的黑色和金色组合为主，辅以一丝白色。但在办公室区，主色调换成了白色。会客室的墙体采用暖金色调，其位于黑色接待厅与走廊后，象征着进入这片区域的中心位置。黑色调则掩藏了真实的"内色"——金色。金色并不眩目，而是缓缓地揭示其存在。

Red Bull Nederland B.V.
红牛荷兰公司

设计师 | Sid Lee Architecture   视觉效果和平面设计 | Sid Lee   项目地点 | NDSM–Plein 26, Neveritaweg 34 1033 WB Amsterdam
摄影 | Ewout Huibers

"To design the inner space, we aimed at retrieving red bull's philosophy, dividing spaces according to their use and spirit, to suggest the idea of the two opposed and complementary hemispheres of the human mind, reason versus intuition, arts versus the industry, dark versus light, the rise of the angel versus the mention of the beast", says jean pelland, lead design architect and senior partner at sid lee architecture.

Inside the shipbuilding factory, with its three adjacent bays, the architects focused on expressing the dichotomy of space, shifting from public spaces to private ones, from black to white and from white to black. our goal in this endeavour was to combine the almost brutal simplicity of an industrial built with red bull's mystical invitation to perform.

The interior architecture with its multiple layers of meaning conveys this dual personality, reminding the user of mountain cliffs one moment and skate board ramps the next.

These triangle-shaped piles, as if ripped off the body of a ship, build up semi-open spaces that can be viewed from below, as niches, or from above, as bridges and mezzanines spanning across space.

In the architecture we offer, nothing is clearly set; all is a matter of perception.

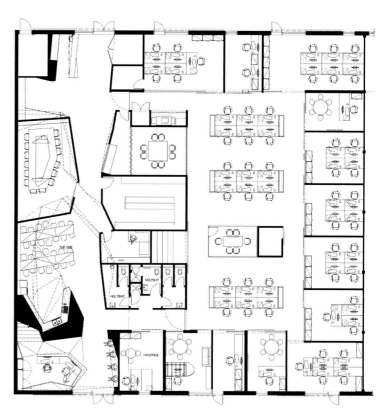

平面布置图

"在室内空间的设计上，我们以唤起红牛哲学为目的，按照功能与精神的方式进行划分，借以人类左右大脑的对立性与互补性为题，提炼出类似理性对抗直觉、艺术对抗工业、黑暗对抗光明、天使崛起对抗野兽传说的观点。"来自sid lee建筑事务所的首席设计师jean pelland。

设计师致力于使空间呈现出一种清晰的功能区域划分，从公共空间到私人空间，由黑到白，再由白到黑。设计团队努力将一个苍白、简略的工业建筑与神秘的"红牛"融于一体并以某种方式进行视觉呈现。

空间多层次的内部架构内涵十分丰富，传递着双重特色，一会儿给用户带来山崖般的视觉感受，一会儿呈滑板梯形。

另外，三角形柱桩仿佛从船身剥落而来，构成一个半开放式空间，从下面看似壁龛，从上面看似桥梁或横跨整个空间的夹层。

总之，在该空间内，一切都不是定制化的，而是一种感知。

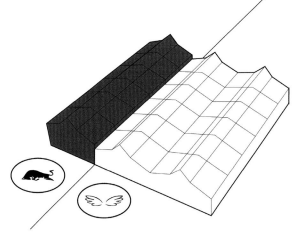

立面图一

立面图二

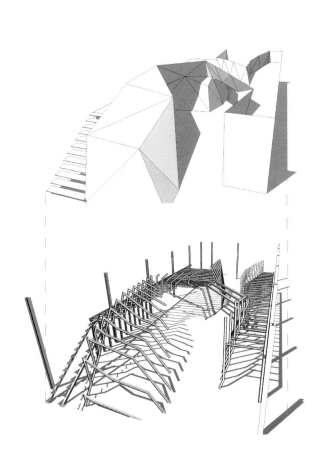

立面图三

LEGO PMD Office
LEGO PMD办公室

设计师 | Rosan Bosch、Rune Fjord   建筑面积 | 2000平方米   项目地点 | 丹麦
主要材料 | 聚氨酯地板、地毯、干燥墙、图形印刷品玻璃隔断墙、吸声天花板、家具   摄影 | Anders Sune Berg

LEGO's designers are the luckiest in the world – they get to play with LEGO all day long! Now, the designers of LEGO's development department, LEGO PMD, has a physical working environment that corresponds to its playful content – a working environment where fun, play and creativity are paramount and where the physical design gives the adults a chance to be part of children's play. With the values 'fun', 'unity', 'creativity' & 'innovation', 'imagination' and 'sustainability' as basis for the design, LEGO PMD has become a unique de-velopment department where the designers can become part of the children's fantasy world. LEGO doesn't just create fun for others – at LEGO PMD working is fun!

In order to create a design that corresponds to the focus on play, innovation and creativity, imagination has been given free rein. Across the room, an existing walkway has been transformed into an oversized sitting environment, where a light-blue padding turns the walkway into a light and soft cloud. The cloud unfolds and expands into sofas, sitting space and a slide that connects the two floors in a fun and playful way. The idea of scale is challenged with design elements such as huge grass wall graphics and a giant LEGO man and tables with built-in bonsai gardens, thus playing with perception and scale – who is big and who is small? Where does work stop and imagination start? Through the physical design, the children's fantasy worlds become part

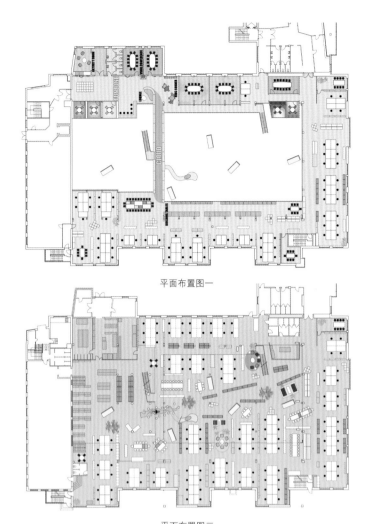

平面布置图一

平面布置图二

of the everyday, creating the setting for the creation of new design for new games and play.

Furthermore, the design of LEGO PMD makes it possible for the designers to work closer together. At ground floor, the open space at the centre of the room creates a dynamic flow where informal meeting pla-ces create a setting for social interaction and exchange of information. Towards the sides, there is room for concentrated work, and specially designed means of exhibition such as the show-off podiums and the model towers give the designers a chance to display their work to each other, facilitating the sharing of knowledge and ideas across the department. On the first floor, an expansion of the balcony has made room for five small and three large meeting rooms in each their own colour with glass facades and a view of the large, open space. A Fun Zone with a yellow table bar creates room for relaxation and social interaction, where a number of building tables for children make it possible for LEGO's youngest employees to test the newest models and products. The new LEGO PMD is the children's universe where imagination reins free – for children and designer alike!

LEGO的设计师是世界上最幸运的,他们可以整天把玩着LEGO。现在,LEGO开发部的设计师们有一个惬意而俏皮的工作环境,在此,兴趣、玩乐和创意是最重要,同时,实物设计给予大人们儿童般玩乐的机会。基于乐趣、团结、创新、富于想象力和可持续发展的设计理念,LEGO PMD已成为一个独特的开发部,这里的设计师们成为儿童乐园的一部分。因此,LEGO不仅为他人创造乐趣,在LEGO PMD工作也是很有趣的。

为了使设计能够集中体现玩乐、创新和创造的主旨,设计师们自由发挥想象力。房间对面,原本的过道被改造成一个超大的客厅,淡蓝色的灯光把过道装点成轻柔的云彩一般,慢慢地舒展开,并延伸到沙发、休息区和用俏皮的方式连接起两层空间的幻灯片区。规模规划面临极大挑战,因为设计元素中包含巨大的草坪墙纸,巨大的LEGO人像,还有带内置的盆景花园的桌子,谁大谁小,从感知和实际尺寸都需仔细斟酌。在实物设计中,儿童般的梦幻世界成为日常生活的一部分,为新游戏和玩具的设计创造了环境。

此外,LEGO PMD的设计可以使设计师们在工作中紧密联系。一楼中央的开放区用于非正式会面,人们可以在这里进行社交互动和信息交流。两边为集中工作区、展览区、模型展示区,设计师们可以在这里展示自己的作品,分享彼此的知识和创意。二楼的扩展阳台上设置了五个小型会议室和三个大型会议室,会议室颜色不一,采用玻璃幕墙以提供开阔的视野。娱乐区的黄色吧台用于休闲和社交,专为儿童设置的桌子供LEGO的年轻员工测试最新模型和产品。新LEGO PMD是孩子们的乐园,在这里,他们可以充分发挥想象力,因此,得到儿童和设计师的一致认可。

ING Bank Slaski corporate department
ING银行斯拉斯科地区企业部

设计师 | Robert Majkut   设计单位 | Robert Majkut Design

The project is based on the concept of free flow of waving walls. Made of two layers, orange and white, they create a kind of stiff curtain freely covering and showing the entrances, passages, and doors in the whole customer zone. The construction of the walls creates the impression of 'internal clothing' being sewed on the fixed building construction – as an additional layer which under its latitude and futuristic geometry hides the ordered, rectangular, harmonized division of rooms. Every hole in the structure shows its thickness and multilayered construction, which is furthermore underlined by internal lighting. The same rule creates the internal island which curves into additional surface, making a kind of landing under the reception, and roofing from which specially designed icicles are hanging – transparent, lighting sticks, shaped into a flowing wave.

When functionality is taken into consideration, solutions of the project are focused on maximum effectiveness. Central, spacious, representative reception is also adjusted for the waiting room with its business TV channels, coffee corner for people who are waiting and also for those who are in the rooms, and the communication network which enables easy access to all offices.

Colours which dominate in the interior are the corporate colours of ING, yet used in much higher intensity than in other branches. Orange is the leading colour in this design, supplemented with big amount of white and few hues of grey.

本案设计是基于波浪形墙壁的概念，自由灵动。一共两层的格局，入口、过道以及整个顾客区门上随意挂着橙色和白色的窗帘。墙壁的设计让人感觉原有的建筑结构穿上了一件贴身的衣服，但又很科学地隐藏在井然有序、四四方方而又和谐统一的房间内。结构上的每一个细节都展现了建筑的厚度和多层化构造，这也通过内部照明作了进一步阐述。内岛按照同样的规则营建，延伸出另外一层，使得接待区下方的地面有了起伏的层次感，而上方的天花板则经过精心打造——透明的光柱形成了流动着的海浪。

　　从功能上考虑，本案注重如何把效益最大化。位于中心区宽敞而独立的接待台也被调整成等候区，商务频道、咖啡台都为等候的顾客而设立，并且设有发达的通话系统，可以轻易连通所有办公室。

　　内部空间的颜色基调就是ING的企业颜色，在总部使用的频次比分行多很多。在该设计中，橙色是主导，同时也穿插着大量的白色和少量的灰色。

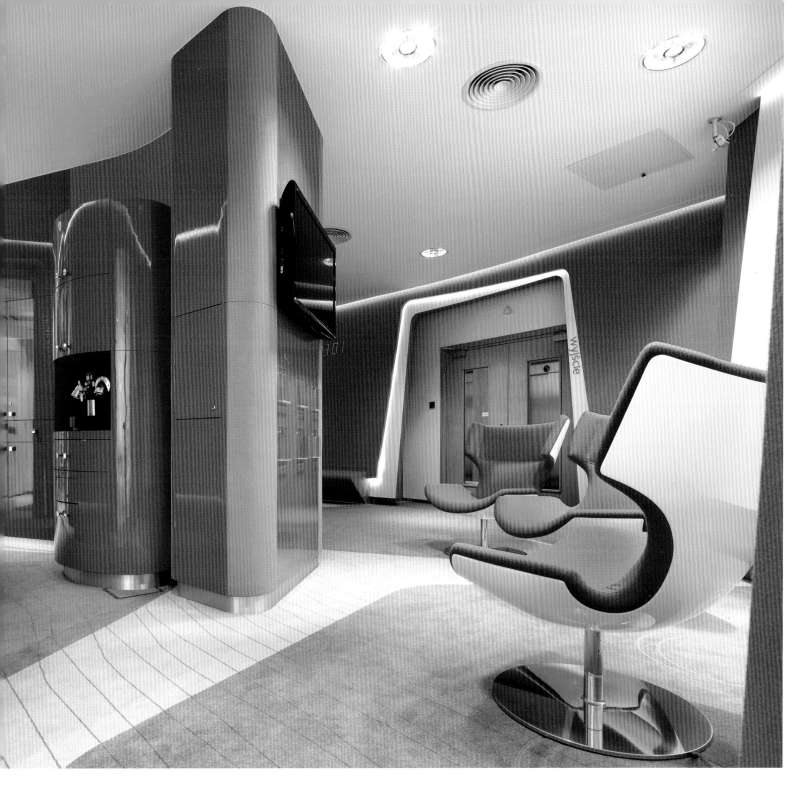

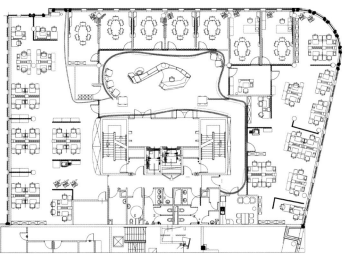

平面布置图一

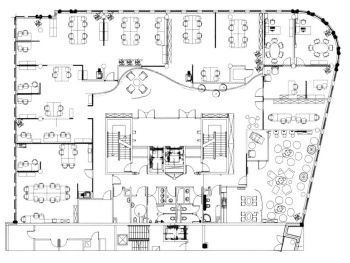

平面布置图二

Zhen Shufen Design Consultance Company Hongkong Office
香港郑树芬设计事务所办公室

设计师 | 郑树芬　　设计单位 | 香港郑树芬设计事务所办公室　　建筑面积 | 250平方米　　项目地点 | 中国香港　　建筑面积 | 250平方米
主要材料 | 灯具、墙纸、瓷砖

Each design firm of Mr. Zheng Shufen has the similar style, but different features, with a philosophy of being harmonious yet different. And his works, whether the big area or the medium area, or the small area, attach great emphasis on the word "balance", such as the combination of hardness, softness and warmness.

The match of soft decorations is also one of the features in the office. The artworks collected by Mr. Zheng in his trips around the world, such as antiques, paintings, sculpture, etc., are dazzling, yet natural and harmonious. They not only provide the office with a relaxed enjoyment, but give the designer the inspirations for the creation.

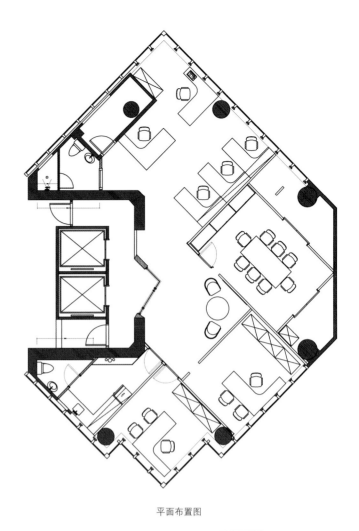

平面布置图

　　郑树芬先生的每一个设计事务所风格都相仿，但又各具特色，有着"和而不同"的理念。他的每一个作品无论是大、中户型，还是小户型，都非常讲究"平衡"二字，比如"硬、软、暖"的相互结合。

　　软装的搭配也是办公室的特色之一，来自郑先生世界各地旅行收藏的艺术品：古玩、字画、雕塑等，琳琅满目却又和谐自然，不仅给办公区带来了轻松愉悦的享受，同样是设计师创作灵感的来源。

Dunmai Office
Dunmai办公室

设计师 | Thomas DARIEL   设计单位 | Dariel Studio   项目地点 | 中国上海
主要材料 | 水泥地面、草坪地毯、乳胶漆、玻璃、马赛克

This project is an office space nested in an old motorcycle factory. It is located in the creative plaza on the South of the Bund, a place that serves as a remembrance to the familiar stories about old Shanghai. The Shanghainese old docks, the tide of the Huangpu River, and the whistles from ships in the morning also acknowledge the nostalgic feeling of the city's ancient past. This location is not only the continuation of the old Bund's elegance and historical prosperity, but also serves as a connection to the newly developed business districts.

After considering all the requests and specificities of the client, a company from Macao organizing creative events, Thomas Dariel decided to construct the office into a pleasurable, relaxing and modern place that reflects the company's dynamism and creativity as well as serving their professional needs. This kind of atmosphere fits to a creative space as well as being functional by using high-tech products.

An internal architectural reconstruction

To memorialize this old 4-floor factory building, the creator kept the building's historical façade, but completely transformed the internal structure into an open, high 3-floor volume arrangement.

This architectural approach opened up the space to obtain brighter and wider rooms, meanwhile revealing the original structure of the building. The internal reconstruction also conveys a friendly atmosphere to the open space required by the client. A 3-floor high

central patio, furnished with large white lacquer desks, was built so that colleagues can easily communicate with each other despite what floor level they are on.

"Work in the Park – Play in the Office"

In terms of interior design and atmosphere, the concept developed in Dunmai is "Work in the Park – Play in the Office."

This concept is reflected in the overall structure and in every detail of this edgy and humorous-looking space.

"Work in the Park"

First, the entering light itself has been designed to pervade the whole space, so that one could feel a sense of being outside. The color scheme consists of pure white in order to change the office's previous image of a dark and old factory into a clean and simple display. As often in our projects, bold colors such as vivid pink or green are used ubiquitously for contrast and for developing natural energy.

Furthermore, the shape of the new internal structure is inspired by the branches of a tree. Drawers on the wall let people imagine that all the plants growing in that space are spreading along the wall and up to the ceiling.

In order to make people feel as if they were surrounded by nature in the given space, the designer attempted to move all the elements of a garden to inside the office: grass lawns under chairs and tables, gardener's tools designed on the walls, swings displayed during afternoon breaks that exhibit the sweet memory of childhood, and a groove for flowerpots on tables.

"Play in the Office"

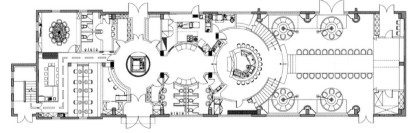

平面布置图一

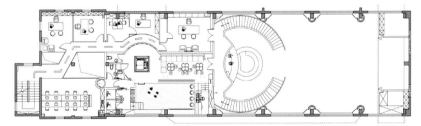

平面布置图二

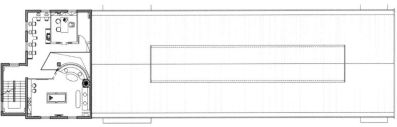

平面布置图三

Playing with interiors by displaying unexpected facilities, misusing objects or creating jokes is part of the Dariel Studio's signature.

In this project, the concept is used to enhance the vitality of the company and to offer a unique and humorous work environment. Dunmai office fully subverts the appearance of a traditional strict office: there are crossroads lining the floors; a tennis court style meeting room with a grid glass wall; floors modeled after blue tennis grounds; and a table set featuring tennis balls.

The restrooms' entrances are designed to look like open elevator doors. Thus, when one is in search for an elevator, he or she will be surprised to find that it is actually a bathroom.

Even the toilet walls are creatively designed, for the designer pays homage to a famous French artist street style by using images from video games to decorate tiled mosaics. The design illustrates that working in an office can be a joyful and unique experience. An office space can be open and transparent, just like the glass walls and doors of many individual spaces in Dunmai Office.

The new design is a reverence to the previous historical building Dunmai Office once was, as well as to the strong heritage and the trademark qualities of the industry's construction style. This creative indoor arrangement makes this space a definite trendy and innovative office, and, at the same time, answers the client's identity and satisfies his requirements.

　　该办公空间原为一个摩托车厂，位于南外滩创意广场，是老上海故事的纪念地。上海老码头、黄浦江潮、早上船上传来的哨声都能引起人们对城市过往岁月的怀旧之情。这个位置不仅是延续老外滩的优雅和历史的繁荣，而且连接起了新开发的商业区。

　　考虑到客户的所有具体要求，来自澳门的公司举办了一系列的创意活动，Thomas Dariel决定将办公室建在一个愉快、轻松而现代的地方，能够反映公司的活力和创造力，也能满足专业需求。该地的氛围正好适合一个创意空间，通过高科技产品的使用也能实现所有的功能需求。

内部重建

　　为了纪念这个古老的四层大楼，创作者保留了建筑物的历史外观，但内部结构完全变更，打造成一个开放式的三层空间。建筑手法上将空间打开以建造更加明亮、宽敞房间，同时也展示出建筑的原有结构。应客户要求，内部重建要在开放空间传递出一种热情友好的氛围。3层高的中央庭院，配有大型白色书桌，以实现在不同楼层工作的员工间的顺畅沟通。

"工作在公园——玩在办公室"

　　就室内设计和氛围而言，Dunmai的理念是"工作在公园——玩在办公室"，此理念体现在这一前卫而幽默的前瞻性空间的整体结构以及每一个细节之处。

"工作在公园"

　　首先，入口处的灯光弥漫着整个空间，给人一种置身室外的感觉。纯白色的配色方案改变了其原本黑暗、古老的工厂形象，变身成一个干净、利落的办公空间。不同于我们以往项目中的大胆配色，如鲜艳的粉红色、绿色的大量使用以形成强烈反差。

　　此外，新的内部结构外形灵感来自树枝。墙壁上的图画让人觉得所有的植物都生长在空间，沿着墙壁蔓延到天花板。为了使人们在该空间中有被大自然包围之感，设计师试图将所有的花园元素搬到办公室，如桌椅下的草坪、墙壁上园丁的工具设计、勾起人们美好童年记忆的秋千、办公桌上的花盆槽。

"玩在办公室"

　　出其不意的设施展示、滥用物件、制造笑话都是Dariel Studio工作室的一部分。

　　在该项目中，设计理念集中于增强公司生命力和提供一个独特幽默的工作环境。Dunmai办公室完全颠覆传统办公室的严肃外观：有交叉设计的楼层间，网球场式的玻璃墙会议室，仿照蓝色网球场地的地板，带网球的办公桌。

　　盥洗室的入口看起来像敞开的电梯门，因此，当人们在找电梯时，或许会惊奇地发现它实际上是一间浴室。

　　甚至厕所墙壁都采用创造性设计，设计师十分崇拜法国著名艺术家的街头风格，所以采用了视频游戏的图像来装饰瓷砖镶嵌图案。设计突出强调在办公室工作也可以是一份快乐、独特的体验。办公空间可以做到公开、透明，犹如Dunmai的个人办公室的玻璃墙和玻璃门一般。

　　新的设计表达了对Dunmai办公室以往历史的崇敬，也展示了工作室深厚的底蕴和行业素养。总之，这一创造性的室内设计带来了一个时尚、充满新意的办公室，同时也满足了客户的需求。

A Bold New Office Environment
大胆的新办公环境

设计单位 | Ellivo Architects' New Design Studio   项目地点 | 澳大利亚布里斯本
摄影 | Scott Burrows (Aperture Photography)

The offices, on the ground level of a high-rise building, are all-black with splashes of yellow and texture. The team was focused on delivering a fit-out that was inviting to staff and clients and representative of how they operate. Utilizing the full volume of the shell of the space revealed the services and workings of the building. Likewise the open plan office is very much on show and clients and consultants are invited to be part of the studio environment.

The new open plan offices have been designed with collaborative ideas form the whole team. The core of the design stemmed from lifecycle and minimalization.

Materials are not used unnecessarily to hide building bones; Arras workstations from Herman Miller have end of life management and no chemical bonding, and the whole layout can be rearranged to suit changing needs under the random zigzag of the suspended lighting.

The new workspace provides a great space for meetings, creativity and collaboration and reflects the young firm's culture and ethos of designing success.

　　本案设在高层建筑的第一层，采用黑色搭配黄色斑点以及纹理图案。设计团队高度关注该建筑对员工和客户的吸引力，广泛听取各方意见，充分利用空间外观来展示大楼内的服务和工作环境。犹如秀台上开放式办公区，客户和顾问被邀请来体验工作室环境。

　　这个全新的开放式办公室的设计思路来自整个团队，设计的核心思路源于生命周期和极小化理念。

　　设计不必使用建筑材料来掩饰建筑构造，赫曼米勒的阿拉斯工作站也有完成使命的时刻，但当光照暂停时，整个格局可以重新安排以适应不断变化的需求。

　　新的工作区为会议、创造活动和协作合作提供了很好的空间场所，体现了新公司的文化和设计的成功。

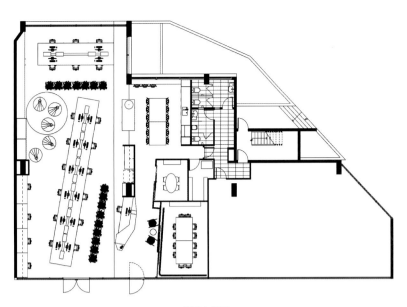

平面布置图

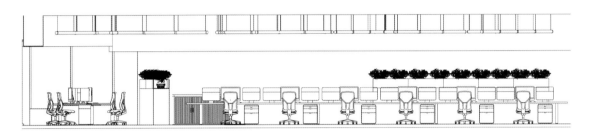

立面图

Forward, Office Extension
Forward公司办公室

设计师 I Shaun Fernandes、Simon Jordan、Markus Nonn、Hanne Gathe   设计单位 I Jump Studios   项目地点 I 英国伦敦
建筑面积 I 900平方米

The reception area was moved from the first to the second floor to improve the internal circulation of the building. Visitors and staff exit a lift into an almost completely red area which feels like a stage set placed in the existing office. A plush carpet and deeply red wooden wall panels provide a strong contrast to the surrounding open plan office which is visible from the reception. Located adjacent to one side of the reception is a small cafe which acts as both holding area for visitors and informal meeting and break out space for staff.

A linear red carpet walkway leads visitors out of reception through the first office space to a black oversized key hole. Slipping through the key hole one arrives in the new office extension.

The first thing you notice here is a gigantic white staircase with a deep red carpet on its steps. On closer inspection the steps reveal themselves to be upholstered seating pads. The open space in front of the staircase is used for larger presentations with the staircase offering seating for audiences of up to 30 people. On a daily basis staff use the stairs as break out or meeting space or to watch their colleagues playing ping pong in front of them.

Next to the staircase a floor length, purple velvet curtain curves around a cluster of social rooms behind: sound booth, billiard parlor and computer games room cater for the staffs entertainment during breaks and after hours. When the curtain is opened up the additional

meeting and break out room directly behind is delineated from the rest of the open space through a chequer board floor which contrast with the traditional grey wood panels on its walls.

The rest of the open space behind the cluster of social rooms is occupied by the canteen. Bar and serveries are all clad in purple-grey wood panels which again feel like staged elements in the otherwise industrial warehouse space. Some of the panels are hinged and conceal secret cabinets for food and drinks. The canteen is open to all staff all day round. At night it turns into a bar for after work drinks and office parties.

接待区从一楼搬到了二楼,以改善大楼内部的空气流通。参观者和工作人员走出电梯,便来到了一个完全是红色的区域,感觉像是设在已有的办公室里的舞台。一块豪华的地毯和深红色的木制墙板与周围开放式办公室形成了对比,办公室在接待区可见。一个小咖啡室在接待区的一旁,既可作为接待参观者的地方,又可作为员工们开非正式会议和休息的区域。

一条红色地毯铺就的走廊引领参观者走出接待区,经过第一个办公空间,来到一个黑色巨大的主要入口。从这个入口穿过,人们就来到了新扩建的办公空间。

来到这里,人们首先注意的就是一个巨大的白色楼梯,楼梯上面铺着深红色地毯。仔细看,那楼梯像软包坐垫一般。楼梯前的空地可以做稍大的演讲,可以为30名观众提供座位。平时,工作人员把这楼梯用作休息和开会的地方,或者是看同事打乒乓球的地方。

楼梯旁,紫色天鹅绒窗帘围绕着一组社交空间:录音棚、台球室和电脑游戏室,供员工们休息和下班时娱乐之用。当窗帘打开时,后面这额外的会议室兼休息室便由格子地板处与那余下的开阔空间划分开来,格子地板与墙上传统的灰木板形成对比。

那组社交室后面余下的开放空间是餐厅。酒吧和餐厅都由紫灰色木板覆盖,在工业仓库里感觉像是舞台元素。一些板子是有铰链的,隐藏了秘密陈列柜,柜里装着食物和饮料。餐厅向全体员工全天开放。晚上餐厅变成了酒吧,员工下班后可以喝上两杯,或者举行聚会。

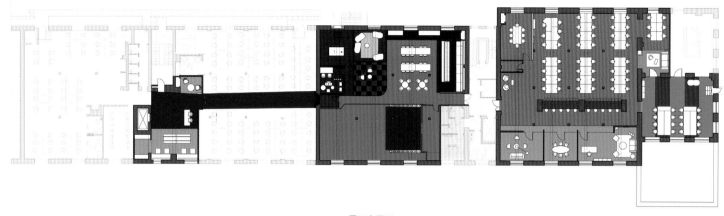

平面布置图

Office BesturenraadBKO
BesturenraadBKO办公室

设计师 | Coen van Ham   设计单位 | COEN! design agency, Eindhoven   建筑面积 | 2200 平方米
摄影 | COEN!, Roy van de Meulengraaf

COEN! created a new working environment and identity layer for the 'Besturenraad / BKO'. These two organizations are going to cooperate more intensively at a new location and take care of two denominational types of education in the Netherlands: Catholic and Protestant. The aim of this project was to visually connect the shared goals and principles of both organizations.

For the design of this story COEN! used The Book as a metaphor. Apart from the Christian and Catholic values a book also consists of structure, text and image. You see stained glass patterns, metal grids based on the golden section and special text prints with a message. The relation between faith and education is also subtly made clear by DNA patterns and golden 'office altars'.

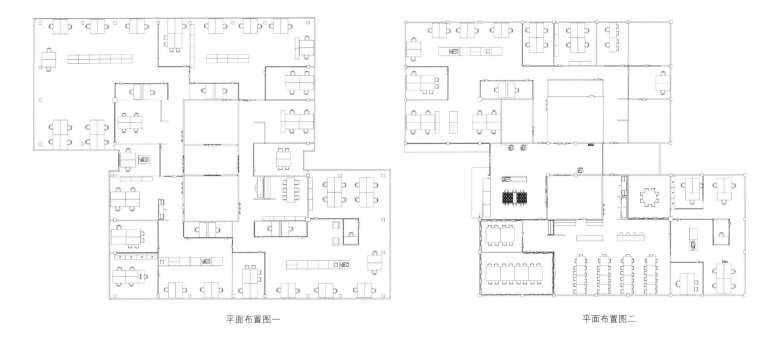

平面布置图一　　　　　　　　　　　　　　　　　　平面布置图二

　　在设计上，COEN!用书作为比喻。除了基督教和天主教的教义之外，书中还需要结构、文本和图片。我们可以看到玻璃纹理的渲染，金属隔栅在金色区域突显，还有标语性质的特殊文本印刷。在DNA和"金色圣坛"的牵引下，游走于信仰和教育之间也是轻松的事。

　　为Besturenarad/BKO营造了全新的工作环境和独特的设计。这两家公司都会在新的工作地点开展更紧密的合作，并致力于两种教义的教学：天主教和新教。该项目的目的是为了让两家组织能更好实现目标并贯穿原则。

San Pablo Corporate Office
San Pablo办公室

设计师 | Juan Carlos Baumgartner、Gabriel Téllez   项目地点 | 墨西哥城Insurgentes Sur   建筑面积 | 3000 平方米
摄影 | Paul Czitrom

Upon entering the space we find an open reception area in a circular form which embraces you as you open the elevator door, this reception area is flanked by two stained-glass windows in blue tones with abstract graphics of medicines and when looked at closely these are infinitely deep. At the center and as the focal point there is a green wall with the logo of the company.

Once you pass through reception you come to a large corridor, which connects the two big rooms for users. As the user explores them they discover playful, colorful and fun spaces that orient them to the use of the entire space.

On entering the large rooms you feel that the space is relaxed, spacious and open, dotted with casual collisions which have their own personalities setting them apart from each other and breaking up the sameness of the work areas. Variation of the pitch and intensity of the lighting generates different atmospheres as you pass through the spaces, lending identity to each area, very subtly framing the function of each space.

Work areas have open vistas, the height of the furnishings contributes to the passage of natural light and all users have exterior views.

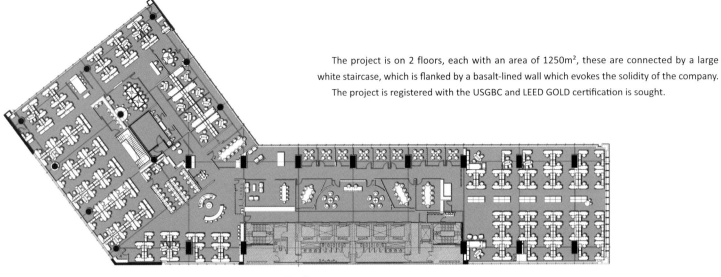

The project is on 2 floors, each with an area of 1250m², these are connected by a large white staircase, which is flanked by a basalt-lined wall which evokes the solidity of the company. The project is registered with the USGBC and LEED GOLD certification is sought.

平面布置图

进入这一空间，映入眼帘的是一个开放式、呈圆形状的接待区。这区域能环抱着刚从电梯门走出的你。接待区两侧装配有蓝调彩色玻璃，玻璃上装饰有抽象图案，当人们走近观察时图案会显现出无限深远的效果。在中心与焦点位置刻有公司的标识。

走过接待区，就来到连接两间容积较大的客户室的大型走廊。用户们可以在此探索这一充满欢愉、色彩、快乐的地方，这处地方能够引导他们享受使用整个空间的乐趣。

当进入大房间时，人们能感受到空间的轻松、宽敞与开放。每一处用于点缀空间的偶然碰撞都有它们独特的、可以区分彼此的个性，这些个性将被用来打破工作区域的雷同感。灯光的亮度、斜度被设计成不同样式，让人们在穿越该区域时能感受到截然不同的氛围。这样的设计能带给每一个空间不同的特性，以及不同的、被巧妙编造好的功能。

工作区域有一个开放的远景。装饰的高度使自然光的借用成为可能，所有的用户也有了更好的外部景观。

项目由两层组成，每一层的面积为1250平方米，由一处用玄武岩装饰边线的大型楼梯相连，这种气派的装饰边线极度彰显出企业的形象。

本项目已在美国绿色建筑委员会注册并通过了LEED GOLD认证。

Xi'an Maiyi Space Design Studio
西安麦一空间设计工作室

设计师丨黄麦一　设计单位丨麦一空间设计工作室　项目地点丨陕西西安　建筑面积丨307平方米

The case is designed for our own ideal office environment, for the Design Studio must have a strong design concept and originality. On the one hand, we enjoy the modern technical and material civilization in this era; on the other hand, we have a sentimental attachment to the past years. Thus, in this case, we create a variable interior space with the integration of Italian simplicity and thick Chinese culture, achieving the combination of the classical and the modern and a concise design style. Different materials well interpret the collision of black and white. Besides, the selection and match of each accessory all reflect the pursuit in details and the high quality of life. On the functional layout, we use hollow partitions, the combination shelves and colorful mosaic wall to divide the space, forming a separated yet continuous office space which brings a more comfortable working environment and better communications.

本案是为完成设计师自身的理想而设计的办公环境，由于是设计工作室就必须具有很强的设计理念与创意。身处这个年代的人们一方面享受着现代的科技和物质文明，另一方面又对过去的年代有着眷恋之情。由此创造出了多变的内部空间，融合意大利式的简约与中式文化的厚重，将古典与现代相结合，以简洁明快的设计风格为主调，用不同的材质演绎着黑与白的碰撞。每一件饰品的选择与搭配都体现着对细节和高端的生活品质的追求。功能布局上利用镂空隔断、组合整体书架以及七彩马赛克墙将空间加以适当区分，形成一个隔而不断、分而不离的办公空间，让工作更加惬意，沟通更加顺畅。

平面布置图

A Dream Private Library in the Urban
都市丛林的梦想藏经阁

设计师 | 张馨   设计单位 | 张馨室内设计事务所/京轩空间设计   建筑面积 | 176平方米

Opening the mysterious door of the library, we can find a space of more than three-meter height. In this space, the designer creates a spacious layout in array type. The long table, chandelier, column care and the base are all adopted the custom or imported large sizes, abandoning the original proportion of the space. Starting from the door, there are plain glossy stones, steady iron parts, gentle gray tiles. Stepping to the heavy long table, the big book cabinet for the collections become the focus of the space. In such a context, some white background gives out a feeling of the cleanliness, especially with the sunlight from the shutters. The Buddha books, Confucian classics, sculpture, crystal, calligraphy, and lights are being in together in a harmonious and leisurely atmosphere. The room is filled with the warmth.

The office takes into account of the elements of the figure of the Buddha and the Heart Sutra. The desk lays against the window, and the imported window-shades brings a sense of stability. The especially custom multi-level desk is to make it convenient for the owners to make a layout according to his habit. Another office continues the calm tone, and the bricks gives a rough original feeling. The decorative ginkgo potted plants also create a Zen mood and help to precipitate mood.

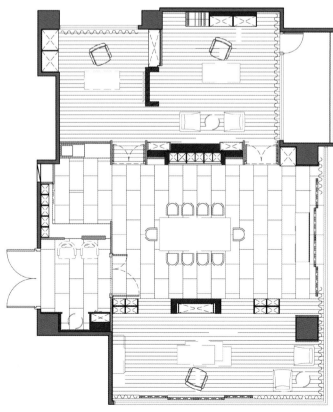

平面布置图

推开藏经阁的神秘之门，超过3米的高度，设计师采用大气、数组式的铺排手法，长桌、吊灯、柱托与基座皆采定制或进口大尺寸规格，放大古堡设定下的空间比例。进入室内，石材削光面朴实，铁艺稳重、灰色地砖和缓；步至厚实长桌，收藏经书的大橱柜与红砖墙成为营造气氛的主要角色，在这样强烈美感的描述下，部份背景刷白衬托出空间色调的干净，尤其在百叶窗引入阳光线条时，佛像、雕塑、水晶、字画、灯影……共同营造出悠闲自在的办公氛围，温暖光氛盈满一室。

办公室预先以佛像、心经定位，背向窗面的办公桌，设计师精心挑选的百叶窗为空间增加沉稳感，设计师特别定制多层次的办公桌，让业主能以习惯的陈列方式布置。另一办公室，延续空间沉稳性，以清水砖堆砌出粗犷的原始感受，银杏盆栽很好地营造出使人沉淀心绪的禅意境。

WirtschaftsBlatt Newsroom in Vienna
维也纳WirtschaftsBlatt新闻中心

设计师 | SANDRA BANFI SKRBEC   项目地点 | 奥地利维也纳   建筑面积 | 1700 平方米   摄影 | Iztok Lemajic

WirtschaftsBlatt is the Austrian leading newspaper which offers the latest economic, finance and financial market news.

In the world where work is increasingly independent of place and time due to modern information and communication technologies, the offices of the newspaper gain a new meaning. The space of the editorial office, which is defined by quick exchange of information, transparency, interaction, a dynamic approach and an innovative spirit, is determined by its final product. Creating the daily newspaper issues and all of its digital applications requires high quality offices in terms of their space and function.

"Change to a new spirit" is the guideline based on which the new office space of WirtschaftsBlatt is designed as a fresh, dynamic and open space, establishing the identity of the newspaper.

The functional concept of offices was dictated in part by the existing space. Therefore, the space is divided into three main zones – the central entrance area, administration part and the editorial office together with production.

The new space organisation creates a connection between open office space and introverted units with various functions (offices, meeting rooms, service rooms, etc.). Half-open elements for meetings,

brainstorming, exchange of information and socialising, are located along the main communication axis, which passes through offices, and they link the space as well as create its visual and sound division.

The organisation and arrangement of each individual part of office space enables a maximum yet controlled flow of information. The open space provides the necessary sound and visual division of individual workplaces by mutual orientation and dependence in the space.

The office space is visually divided by subtle graphics, which link different elements into a whole at the same time. The graphic elements represent the iconised contents of WirtschaftsBlatt with the typical identification elements of Vienna and Austria.

Furniture design visually and functionally depends on both the existing space and identity of the newspaper. Its subtle design and mathematical arrangement enable transparency and openness of the space with a great deal of privacy of an individual workplace, especially in open office space. Colour emphasis establishes the diversity of the space and the connection with the newspaper's identity. Each individual zone in office space includes specially designed elements which follow their function and establish the central points of events due to their design, thereby dictating the dynamics and flow of the space.

Through the demanding implementation, which is reflected in a consistent architectural and functional concept in new offices, an important role was played by the austrian company M.O.O.CON GmbH.

The result of excellent cooperation in the project between the client and architects is an office space which embodies WirtschaftsBlatt with its architectural design, the design of furniture, the colour concept and graphics.

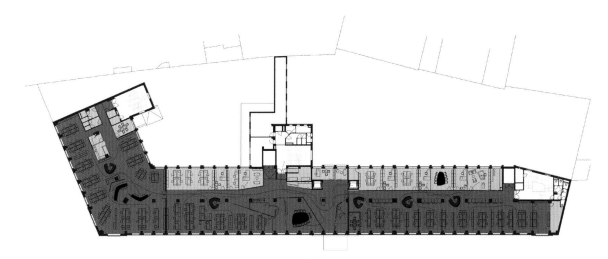

平面布置图

WirtschaftsBlatt是奥地利的主要报纸，给人们提供最新的经济、金融及金融市场信息。

由于现代信息和通信技术的运用，人们工作的时间和地点日益独立，因此，报纸的办事处也有了新的含义。编辑部办公室要求实现信息快速交换，透明度好，互动性强，具有动态方法和创新的精神，因其要带来最终成品，日报的制作和所有数字应用对办公空间和功能都有很高的质量要求。

"转换成一种新思路"是此次WirtschaftsBlatt办公空间设计的出发点，打造一个充满活力、动态的开放空间，树立起该报纸的社会地位。

办公室的功能理念在某种程度上决定现有的空间尺度。因此，空间被分为三个主要区域，分别为中央入口区、行政管理部、编辑制作部。

新的空间构造建立起了开放式办公空间与封闭办公区的连接，拥有各种功能，如办公室、会议室、服务室等。用于会议、集思广益、信息交流和社交的半开放式区域沿主要通道设置，通道通往所有的办公室，连接起各空间的同时也带来视觉和声音的划分。

每一个办公空间都实现了最大而有序的信息流量。开放空间为单独的工作场所提供了必要的声音和视觉功能。

办公空间看似是按图形划分，其实这些图形又将不同的元素连接成一个整体。在典型的维也纳和奥地利的识别元素映衬下，图形元素代表着WirtschaftsBlatt的标志性内容。

家具设计从视觉和功能上都依赖于现有的空间和报纸的地位。巧妙的设计和数字设置使独立的工作场所拥有隐密性的同时又具备透明度和公开性，尤其是在开放的办公空间。对色彩的强调突出了空间的多样性和报纸的特殊地位。个人办公空间按照其功能有独特的设计元素，有中心聚集点，也有空间动态和流动感。

一致的架构和功能理念在新办公室设计得到严格的执行，可以说奥地利M.O.O.CON GmbH公司在这方面充当了重要的角色。

在此项目中，客户和建筑事务所有着良好的合作，通过建筑设计、家具设计、颜色和图形的运用，共同打造WirtschaftsBlatt的办公大楼。

Noble Bank
莱宝银行

设计单位 | Robert Majkut Design   摄影 | Olo Studio，Lukasz Krol

Noble Bank is an exceptional financial brand, specializing in private banking. Noble Bank combines the knowledge and professionalism of their personal advisors with its unique, tailor-made financial products.

The inspiration for this project came from an old English interior bank design and photographs from the mid 1920's, which evoked the tradition and history of those days.

The interior creates an impression of timeless elegance and prestige. The reception area, conference room, and corridor wall facings and shining veneer plate are in dark, chocolate-like colors, the ceilings are made of upholstered dark brown textile panels. There are elements of cornices and columns finished with veneer. Recessed lighting was used in the ceiling; however, individually designed table lamps as well as a limestone floor were emphasized.

Private offices were specially designed for individual meetings with clients. Each room is in a different color but together they create a flowing unity defined by tones such as warm beige, cream, gold, and sky blue. They are equipped with round black glossy tables and four armchairs covered with sky blue suede. A unique floral pattern was created for wallpapers, glass wall divider between the reception area and the conference room, which gives the impression of harmony of

the entire design. Noble Bank starts a new trend – the return to the retro style – and brings it to a new level where it meets modern design.

The entire aesthetic solution appeals to people who admire high quality, professionalism, and aesthetic perfection. This is a place for clients who expect supreme level of service in a financial institution, no different from what is established in London, New York, Zurich, or Monaco. Noble Bank offers to its clients the same quality, comfort and high quality services.

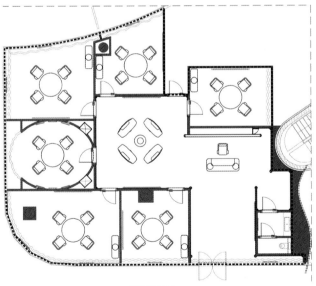

平面布置图一

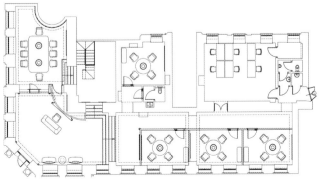

平面布置图二

　　莱宝银行是一个特殊的金融品牌，专门从事私人银行业务。莱宝银行将个人顾问的专业知识与其独特、量身定制的金融产品相结合。

　　该项目的灵感来自20年代中期一家古老的英国银行的室内设计和照片，这让人想起传统和过往的那些日子。

　　内饰打造了一个优雅和充满信誉的形象。接待区、会议室、走廊壁饰和闪闪发光的木饰面采用黑色和巧克力色，天花板使用了深褐色软垫纺织板，还有飞檐和木饰面柱子。另外，天花板采用了隐藏式照明，强调突出了专门设计的台灯以及石灰石地板。

　　私人办公室专门用于与客户的单独会见。每个房间采用不同的颜色，但它们一起形成了流动的统一色调，如米色、奶油色、金色、天蓝色调。所有会议室都配有黑色圆形桌和四把天蓝色绒布面扶手椅。独特的花卉图案壁纸装点着接待区和会议室墙壁，整个设计给人一种和谐之感。莱宝银行开创了一个新趋势，即回归复古风格，与现代设计的融合将其带到一个新的高度。

　　人们在设计上追求的高品质、专业化和高审美度，这里一一呈现。同时，这里也为客户提供最高水平的金融服务，丝毫不差于伦敦、纽约、苏黎世或摩纳哥的银行。莱宝银行向客户提供同等质量、舒适和优质的服务。

Open Finance
开放式金融

设计单位 | Robert Majkut Design 摄影 | Olo Studio

As the symbol of caring about the customers the orchid flower was chosen for the project leitmotif. Orchid embodies protection and attention to unique needs of every single person. It is a metaphor of individual answer to client's expectations. Afterwards, full standardization of network has emerged. The branches were comparable in terms of size and number of customer service points. Modern materials and light coloring of interior were used. White walls contrast with the elegant orange chairs for clients and staff, and with the violet of some furnishing elements. Light curtains made of material with soft, smooth lines give the feeling of privacy. Thanks to eye-catching lightning, emanating optimism and accessibility, branches are easily recognizable also in the evening and at night. Thus, through design, a completely new value was created – an image of a place corresponding with the character and company values of Open Finance.

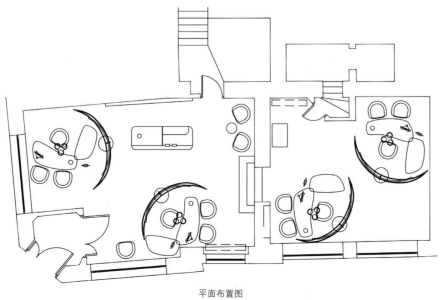

平面布置图

从关怀客户的角度出发，兰花被选为项目的吉祥花，体现出对每一个人独特需求的关注和保护。这也是对客户期望的解答，随后，网络已完全标准化。在规模和服务点的数量上，分公司都是均衡的。现代装饰的材料和室内光线色料都被使用。雪白的墙壁映衬着橘色椅子，同时还搭配着紫色的家具渲染。轻盈的窗帘由质地柔软、光滑的面料制成，很好地保护了隐私。多亏了有绚烂夺目的光线，使人感觉乐观顺畅，各部门也日夜可辨。因此，该设计创造了一个全新的价值理念——一个能体现开放式金融价值和特色的梦幻地方。

Le Cube Office
乐立方办公空间

设计师 | 谢利俊　　设计单位 | 汕头纳天优联设计机构　　建筑面积 | 680平方米　　摄影 | 邱小雄

This case is for a famous brand and with the basic principal of "mainly focus on design but not decoration", and style of "simple and pure". This comes into being the personality of the office which can make workmates happy and client impressive.

Both the image painting at the entrance and the eye-catching horizontal logo are fashionable but not luxurious. The charm of the brand is presented to the visitors through declined passenger and irregular light boxes. The right side is half-open living room, and then tea and coffee place. Black and white newspaper and wallpaper, pre-kept photo walls can present humane and harmonious enterprise culture in a leisure way. Two or three workmates can play table games or read "Weekly News" with a cup of coffee, which is a so pleasant thing at tea break time. The leisure area is accessible to toilet and meeting area, can also to other areas. This design cannot only avoid the interruption brought by the passers-by, but also enlarge the spatial feeling and keep the flexibility of the space.

The unique design in the corridor is with colorful organic glass and stainless square tube as partition at one side, changeable and fashionable, while with white wall at the other side, natural and simple. Two different materials are used in the same corridor,

Turkcell Maltepe Plaza Office
TURKCELL MALTEPE 广场办公室

设计师 | Ayca AKKAYA KUL & Onder KUL　　设计单位 | mimaristudio　　建筑面积 | 2100平方米
项目地点 | 土耳其伊斯坦布尔　　摄影师 | Gurkan AKAY

Turkcell Maltepe Plaza renovation project was led by Turkcell Communication Services Building Managment Department and designed and executed by mimaristudio in a total of 4 months.

Main objective of the project was to provide a new and desirable social/communal areas to improve work conditions and wellbeings of the highly educated young professional employees.

2.100m² of works have been carried out in a building with a high end technolgy infrastructure in overtime and in three phases; in the administrative, South and North wings.

The first phase of the project consisted of renovation of the fitness room, locker rooms and public toilets of the adminstrative wing. The second phase consisted of renovation of the break out rooms, coffee station areas, interior landscape areas, toilets and the main restaurant on 4 floors of the South and North wings.

Building management and employees of each department were interviewed early on to help develop unique designs for each department's needs. A common language of materials has been developed and used throughout the project while variations in accents and loose furniture depict each department's unique needs.

All communal spaces break out areas and coffee stations, formerly treated as back of house spaces, have been integrated in to the open workspaces with the help of same materials used throughout.

    2 existing atriums were also lanscpaed to be in unison with the new 8 communal spaces on 4 floors. Furniture, lighting design, materials and the new interior layout of the busy restaurant provides a cosy athmospehere for the employees. Public toilets in the premises have been designed and furnished only with functionality in mind.

    Mimaristudio's overall intent was to provide multifunctional design elements, such as very specific dining furniture to allow for formal/informal and spontaneous gatherings, breakouts, meetings, hosting guests, banquets and dining in the newly formed communal areas.

    Mimaristudio custom designed and furnished all millwork, specified all loose furniture and accessories.

    In accordance with the owner's company policies, environmentally friendly materials and equipment were used throughout the project.

    Lighting design, led by mimaristudio, was executed by Dark Lighting with the help of photometric and proper selection of light fixtures, to provide adequate lighting in each area of work.

    Final product proved to be successful as commended by both the employees and the executives of the Turkcell Group of Companies.

Turkcell Maltepe广场改造工程设在Turkcell通信服务大厦,mimaristudio通过4个月的时间来设计。

该项目的主要目标是提供新型而又可实现的社会或者公共领域来改善受过良好教育的年轻雇员的工作条件和福利。

2100平方米的工程在一座拥有高端技术的大厦内实施,包括三个阶段的在建技术基础设施:行政楼层和南北两翼。

项目的第一阶段包括健身房、更衣室和行政区旁边的公厕改造。第二阶段包括休息客房、咖啡区、室内景观区、厕所和4个楼层南北两翼的主餐厅。

在早期的时候,设计团队和各部门的员工进行了面谈,更好地为各部门量身设计。在整个项目中,一种通用的材料语言已开发和使用,但在家具的变化上仍然可以体现出各部门的独特需求。

所有的公共空间都打破了地域和咖啡区的限制,之前整个房子后部空间应经用相同的材料打造成工作区。

两个中庭通过造景,与4个楼层的8个公共空间形成整体。家具、照明设计、材料和生意兴隆的餐厅内新的布局,为员工提供了温馨的氛围。楼宇的公共厕所设计和装饰也十分贴心。

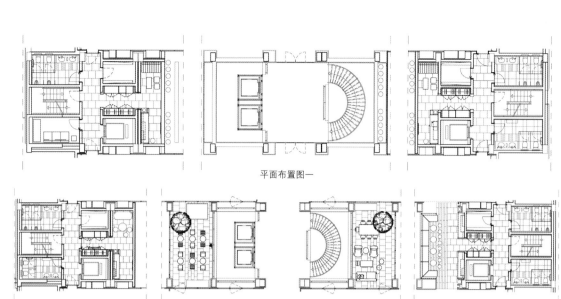

平面布置图一

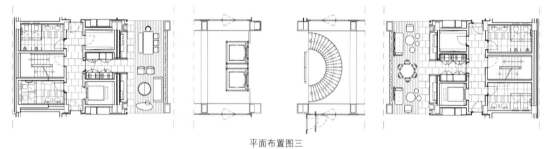

平面布置图二

平面布置图三

mimaristudio的整体想法是提供多功能的设计元素，例如，把非常独特的餐厅家具用在正式或者非正式和自发的聚会、休息、会议、招待客户、宴会和新成立的公共区域就餐。

Mimaristudio根据客户需求，定制和布置所有木制品，确定所有软装的家具及配件。

按照客户公司的政策，在整个项目中使用环保材料和设备。

由mimaristudio牵头，达科照明公司进行光度测量和灯具的选择，在工作区域的各个角落提供足够的照明。

本案最终获得了Turkcell公司集团高管及员工的一致认可和赞誉。

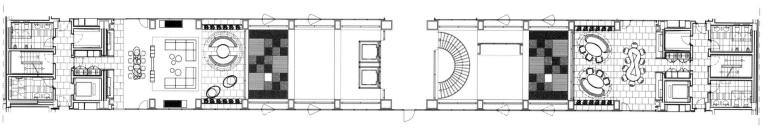

平面布置图四

Ymedia
Y媒介

设计单位 | Stone Designs 建筑面积 | 1200平方米

The Ymedia Project has represented a major professional challenge for Stone Designs.

It has been that way because of enormous space, 1200 sq. m, shared by three different companies, each of which with their own specific needs.

Maybe the fact of conceptualizing it while we were in Tokyo had a great influece of Japanese values and habits on the project, so we did not focus so much in the global space but in the well-being of the people who would work there.

The objetive was making the people that work in Ymedia feeling as they were at home. The employees well-being has a reflex in the office good working.

We have used a small range of materials in order to obtain a simple and coherent view. Both the furniture and the interior design try to bring landscapes to our minds, as in the case of the table dividers, which recreate a sunset in the horizon.

The oak tree has been the main material, and it has been mixed with different and intense colored fabrics. Stone Designs has tried to take nature into Ymedia offices.

In short, we were looking for a revolution in the office relationships, creating a real, authentic and excellent work atmosphere.

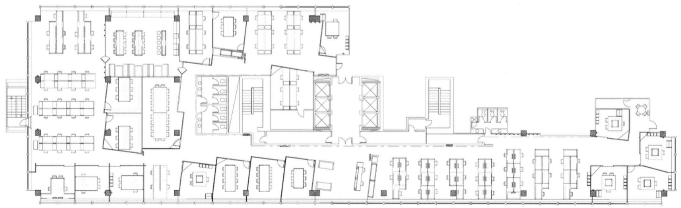

平面布置图

对设计师而言，"Y媒介"项目意味着一次巨大的专业挑战的来临。因为在这面积达1200平方米的领域内，有三家完全不同的公司，这三家公司对设计的需求更是不尽相同。

也许因为设计师身处日本，深受日本文化和习俗的影响而倾向于将项目做概念化调整，所以相较于环球空间，设计师将注意力更多地放在人们在此工作时如何享受这个空间这一命题上来。

让人们在这里工作就如同呆在家里一般，员工们就会更好地投入到办公室事务上去。

设计师尽量使用较少类型的材料旨在获取一个简单、紧凑的视觉感受。家具与内饰都给员工的思维呈现出一幅幅风景之图。例如桌面分隔板的营造给予员工一个太阳从地平面升起的感觉。橡木是设计中的主打材质，其融合了诸多颜色的纤维材料。设计师运用橡木的自然美，将办公室的环境设计成天然的大自然景观。

本案设计师致力于给职场带来一场新观念的革命，营造一个真实、优越的工作氛围。

XYI OFFICE DESIGN
大隐于市的设计

设 计 师 | 黄士华、袁筱媛、孟羿彣　设计单位 | 隐巷设计顾问有限公司　项目地点 | 中国台北　建筑面积 | 100平方米
主要材料 | 水泥板模墙、外墙质感涂料、黑铁板、拉丝不锈钢、木色夹板、桧木板、染色桧木板、大甘木木皮、白色密度烤漆板、PVC、强化玻璃、黑色强化玻璃、灰镜、水性白色烤漆、人造石、银狐理石、复古面印度黑理石、金属砖、岩砖　摄影 | 王基守

The Ymedia Project has represented a major professional challenge for Stone Designs.

It has been that way because of enormous space, 1200 sq. m, shared by three different companies, each of which with their own specific needs.

Maybe the fact of conceptualizing it while we were in Tokyo had a great influece of Japanese values and habits on the project, so we did not focus so much in the global space but in the well-being of the people who would work there.

The objetive was making the people that work in Ymedia feeling as they were at home. The employees well-being has a reflex in the office good working.

We have used a small range of materials in order to obtain a simple and coherent view. Both the furniture and the interior design try to bring landscapes to our minds, as in the case of the table dividers, which recreate a sunset in the horizon.

The oak tree has been the main material, and it has been mixed with different and intense colored fabrics. Stone Designs has tried to take nature into Ymedia offices.

In short, we were looking for a revolution in the office relationships, creating a real, authentic and excellent work atmosphere.The Ymedia Project has represented a major professional challenge for Stone

Designs.

It has been that way because of enormous space, 1200 sq. m, shared by three different companies, each of which with their own specific needs.

Maybe the fact of conceptualizing it while we were in Tokyo had a great influece of Japanese values and habits on the project, so we did not focus so much in the global space but in the well-being of the people who would work there.

The objetive was making the people that work in Ymedia feeling as they were at home. The employees well-being has a reflex in the office good working.

We have used a small range of materials in order to obtain a simple and coherent view. Both the furniture and the interior design try to bring landscapes to our minds, as in the case of the table dividers, which recreate a sunset in the horizon.

The oak tree has been the main material, and it has been mixed with different and intense colored fabrics. Stone Designs has tried to take nature into Ymedia offices.

In short, we were looking for a revolution in the office relationships, creating a real, authentic and excellent work atmosphere.

台北市的巷弄里卧虎藏龙，有许多不起眼甚至可能隐密到你不知如何上门的商家，而我们的想法是希望表达如"隐巷"字面般，一种低调、实务、质朴的理念，一种柳暗花明又一村的创新。

　　基地是位于台北市大安区巷弄里的旧民宅，六七十年代纵长向的建筑，前为小于3米的巷弄，后为1.5米的防火巷，采光差；但我们喜欢那样的年代，喜欢时光停留的错觉。寸土寸金的台北市区，大部分人总是尽可能地放大住宅功能空间，我们拆掉旧有车库后，留出前院空间，置入景观植栽与生态，让转折的巷弄与密集的住宅呼吸，而公司同事上下班时能过渡转换心情，门前的塑形鸡蛋花、会议室外的白水木与生态池让夏天的台北绿意盎然，让冬天的台北散发禅意，搭配门上桧木的淡香，使材料本身就是一种设计语汇；拆除旧有雨遮，仅保留会议室的部分，新的外露I字形钢作为结构梁，支撑会议室屋顶，我们去工厂采购了剩余的废弃桧木料，切成50毫米宽的木片，以45°的方式拼贴，这是公司标志的夹角角度，也向80年代流行的拼贴设计致上敬意，使设计回归到手工与材料质感，并非仅是追求创新。

　　室外墙面材料使用板模灌浆，主要是为了做防水处理，并赋予设计语汇，右斜的墙线与左斜的会议室玻璃，如果你站在路口隐约能看见X的交叉点，入口地面使用黑色复古面理石用来定义室内外空间的界线，建筑原有结构柱采用黑色铁板包覆，使其氧化减慢，契合那旧建筑的年代；会议室外墙采用强化玻璃，解决室内采光不足的问题，也是节能方式。

　　内部我们保留当初拆除打凿的痕迹，这是建筑生命的周期呈现，与新的材料产生冲突感，却同时互为搭配，依据功能区分共为六区，会议室与员工训练区、Mini Bar互为重叠，电视嵌入材料柜门中，使其能90°旋转，根据需求使用，会议桌由大甘木、玻璃与黑铁结合而成，搭配意大利品牌单椅，此区同时有塑料、镜面不锈钢、玻璃、木夹板、黑铁板与水泥相互冲突产生的协调空间，而书柜从平面也采用斜度处理，Mini Bar内置入热水机与过滤器，使用Kohler厨房龙头，三节式的设计让机能更臻完整，上方镜面壁柜延伸至室外，在玻璃隔间区域转换为黑铁板，一是为增加室内外的连接，二是隐藏墙电箱，黑铁与板模墙面呈现出建筑Rough texture。

　　整体设计我们希望能呈现材料本身的质感，无论是平整的墙面与粗糙的铁板还是手工质感的木板与镜面玻璃，透过比例将原本产生许多冲突的空间与材料转化成空间里的主角，低调却是空间的生命。

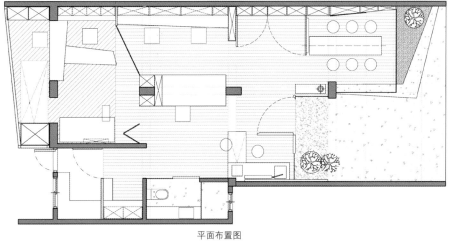

平面布置图

Youth Republic
青年王国

设计单位 I KONTRA　　设计团队 I Cem Demirturk、Gulsah Cantas、Pelin Peker、Elif Baltaoglu、Sedef Gulenler
平面设计 I Monroe Istanbul　　项目地点 I 土耳其伊斯坦布尔　　建筑面积 I 1300平方米　　摄影 I Onur Solak

From an old atelier building to a Youth Republic... This 1300m² of ancient space is now a contemporary loft for a youth agency with nine departments. Situated right in the middle of the trade center of Istanbul, this visual transformation was authored by KONTRA, one of the most prestigious interior architecture firms based in Istanbul.

The entrance welcomes guests with a 6 m high ceiling and gray rusty metal plates. On them, the philosophy of the company is laid out. This mature masculine appearance emphasizes the strength of the young team inside.

Inside, the lively and colorful space screams the inspiration for the young team: "Stay Young!" This is an ideal environment to encourage creativity, concentration and relaxation. The first thing that draws attention, is the nuclear reactor shaped meeting point right in the middle of the open office. It includes a heritage table – a witness of the creative mind of this young agency since its foundation. This space is the heart of the Youth Republic and is used for brainstorming.

The entire space is designed as an open office. It makes you feel the spirit of a university campus with its coloful lockers. Furthermore there are urgent meeting points where the team will be able to jot down their ideas on blackboards.

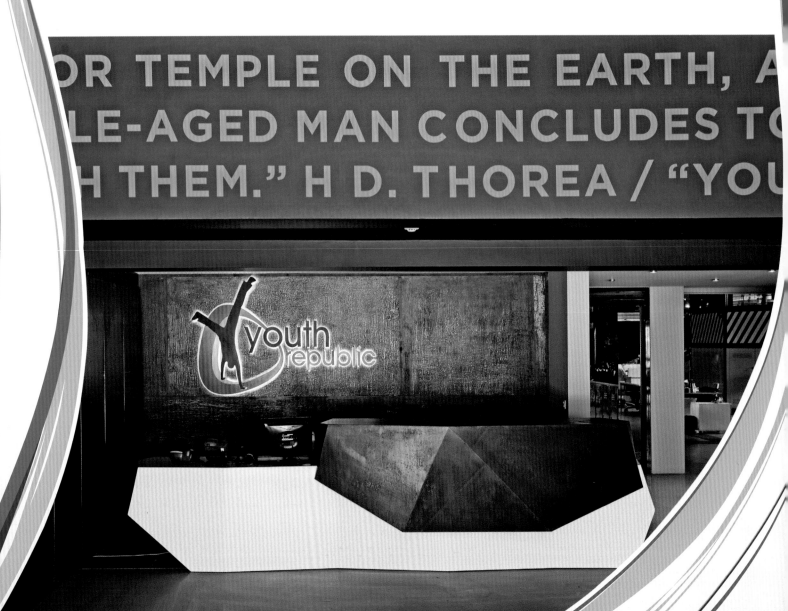

Using raw material reflects the company's work principle and young team. The agency itself transforms abstract ideas into concrete ones.

The tangarine trees between the tables in the open office provide fresh air for the employees as well as creating a refreshing atmosphere. The canteen is the point of discharge for the employees during a busy day of brainstorming. Fun images on the walls and colorful fluorescent lights on the ceiling maintain the young energy of the office. The yellow striped walls are calling on the employees to "take a break".

Almost all parts; meeting rooms, training room, even managers' rooms are designed as a semi-open space. The glass seperation units with graphics support the energy circulation between the managers and the team.

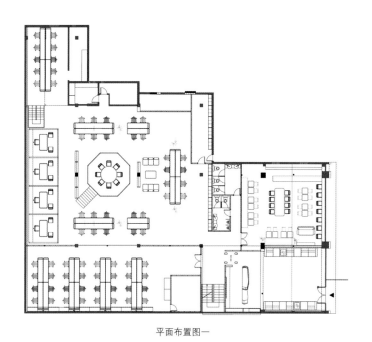

平面布置图一

平面布置图二

  从老旧工作室变身成一个青年王国,这个1300平方米的古老空间现在是一个充满当代感的顶层空间,拥有九个部门的青年机构。作为伊斯坦布尔最负盛名的室内建筑公司之一,KONTRA特为这个坐落在伊斯坦布尔贸易中心中央的建筑打造视觉变换效果。

  入口处也是迎宾区,由6米高的天花板和灰色金属板构成,上面印着公司的经营理念。另外,成熟而充满阳刚气质的外观突出了年轻团队的力量。

  生动活泼的室内空间有效地帮助激发年轻团队的灵感,正是"永葆青春"的写照。理想的环境有助于带来创造力、凝聚力和舒适感。进入空间,首先映入眼帘的是位于开放式空间中央的核反应堆形接待区,包括一张流传下来的桌子,它见证了这个年轻机构自成立以来的创造性思维,因为这个空间专门用于头脑风暴。

  开放式办公空间犹如大学校园,里面是多彩的储物柜。此外,紧急会议的黑板可以让员工记录他们的想法。

  原料的使用体现了该公司的工作原理和年轻的团队,很好地将抽象化概念具体化。开放式办公室内的蜜橘树带来了新鲜空气,也给员工创造了一个清爽的氛围。食堂是员工在忙碌的工作期间放松的好去处,墙壁上的图案和天花板上多彩的荧光灯保持了办公室的年轻活力,同时,黄色条纹墙壁似在呼吁让员工"休息一下"。

  空间其他部分,如会议室、培训室,甚至经理室都采用半开放式设计。图形玻璃隔断允许经理和团队之间进行有效沟通。

Splendid Shows for an Antiquated Space
大木和石loft空间

设计师 | 陈杰　　设计单位 | 福州大木和石设计联合会馆　　项目地点 | 中国福州
建筑面积 | 585平方米　　摄影 | 周跃东

The office of Big Wood and Stone Designing Group was decedent from the store of a beer brewery's and restored with new meanings by the designers. Whatever is the 3-dimensions visual sight or is the scene filled with the elements of Chinese or Western's, are all full of the things that worthy people's thinkings. The insists the designers have on the art is just like the energy the unique architecture owns. It diffuses the individual's aroma, with no bounds that comes from the time and the space.

Although the office is located at the urban area, the designer has successfully combined the natural scene and the indoor environment. Therefore, the office could be kept distance from the noisy metropolis. A quiet and mysterious atmosphere brings a sense of peace to the antiquated and modernized space. The 300 square meter-sized old store is nearly 10 meter high. The height makes the space dreamful and give people an eager for touching the sky. Because the designers have to keep the soul of industry time and have to make innovation on it so that they can give a new start to the building, the project introduced Loft Style to here. It means a challenge to Chen Jie and it is a reconsideration for the designs.

The inner space was divided into several levels regarding to the

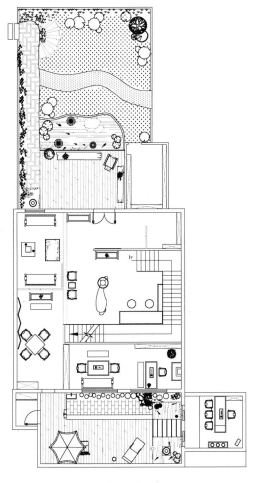

平面布置图

height. The old affection for the brewery was not vanished. Meanwhile, the building's frame becomes more interesting. People living in different floor can see each other. You might felt incredible due to the good control of the designer's on the space's structure. Besides, the transition created by the lights gives the space a dramatical content. The bar area on the left side is a visual hotspot because of the wooden vine shelf and the steady aroma diffused by the table. You would find a sense of root-took while standing here. The lights and shadows infiltrate the shutter to communicate with the space of background. All of them are expressed through a simplified complexion.

The designers have humbly grew the new building on the base of the old one with the understandings of their own. The concusses made during the progress caught our sights and compliments. Saving, respects, salute, reconstruction and rebirth, the new office seems to perform a drama both for old days and nowadays.

　　大木和石空间设计的工作室是由啤酒厂仓库改建成的建筑，经过设计师的提炼再造，焕发出新的光辉和意义。无论是立体错落的视线，还是亦中亦西的布景，都有道不尽的耐人寻味在里面。设计师对于艺术的执着追求，正如这座独特的建筑本身蕴藏的能量，它洋溢着个性，没有时间的限定，没有空间的约束。

　　大木和石空间设计的工作室虽置身闹市区，但设计师将自然景观与室内环境相交融，使其远离喧嚣，一种静谧的氛围为这古朴而又兼具现代感的空间平添了几许淡定。这座三百多平方米的旧仓库有着近十米的层高，这样的高度使这里如同梦境一般，让人产生要与天际无限接近的渴望。LOFT风格是这里的标签，既要忠于工业时代存留下的精髓，又要对其进行再创造，赋予建筑以崭新的生命力。这对设计师是个挑战，更是一次对设计的重新审视。

　　内部空间在高度上被隔出了几个层次，老厂房的情感仍在，建筑结构却有趣了许多。而身处不同楼层之间的人们，可以观望到彼此的身影，上下楼梯间，你会感叹设计师对空间结构的良好掌控。此外，设计师通过灯光营造出的明暗过渡，则让空间产生了丰富的戏剧性。左侧的吧台区是这里的一个视觉焦点，厚实原木搭建出的酒架与台面释放着沉稳的气质，置身其间，人有种扎根的感觉。百页窗拉出的光影则与背后的空间有了交流的可能，它们都以一种朴素的情怀表达出来。

　　设计师用自己的理解让新建筑以谦逊的方式在旧建筑上继续成长，这一过程中的碰撞吸引了我们的目光，并为之称赞。存留、致敬、改造、重生，大木和石空间设计的工作室仿佛一场时空交错的舞台剧，演绎着过去与当下的精彩。

798 Art Factory
798艺术工厂

设计师 | 戴国军　设计单位 | 北京壹柒设计有限公司　项目地点 | 中国北京　建筑面积 | 280平方米
主要材料 | 乳胶漆、水泥自流平、木板

The thinking to remove the numbers makes clear the measurements and physical variations in the relative space. Among the original invisible measurements, the unremovable ones affected people' life. The duties of the designers' are combining and using the measurements which are hiding in the space to return to the natural start point.

The designer adjusted the number of the moving lines to the least. No matter the iron slabs, reinforcements, wood ceilings, iron structures or glass, the designer can easily melt the essences of the materials together and introduce the history and cultural elements to the space.

　　归零的思考，能理清相对空间的尺度与物理量。在原有看不见的尺度中，固定的尺寸影响着人们的生活，设计师的天职就是组合并擅用这些隐藏于空间的尺度于人易于使用，使人舒适，创造生活的机能，回归自然的原点。

　　设计师把动线调整到最少，贯穿全区。不论是铁板地坪、清水混凝土、实木吊顶，还是钢结构和玻璃，将这些材料的本质巧妙地融合，再综合其历史文化充盈整个空间。

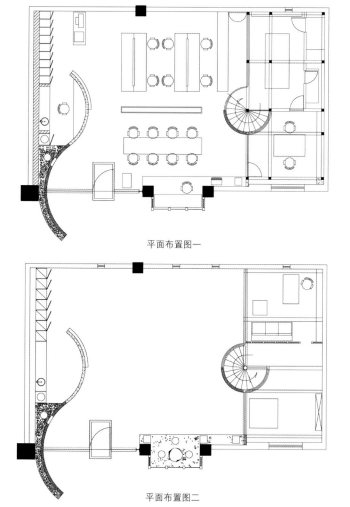

平面布置图一

平面布置图二

Decoding DNA
解码DNA

设计师 | 陈方晓   设计单位 | 陈方晓设计师事务所

Design concept originated in the owners of business services: for each different company's core operations process design computer programs, just as DNA coding.

Then DNA was symboliced, decoding the main context for the design, throughout the entire space to finish.

DNA decoding the image on the wall for the reception hall of the art equipment - the owners and their customers LOGO profound feelings.

DNA decoding for the soft lighting

DNA decoding office to facilitate the practical console, a connection to another office tour office tour will be long and narrow office space well be reduced to an organic Entirety.

设计概念起源于业主企业的服务内容：为各个不同企业的核心运营过程设计电脑程式，就如同DNA编码。

于是DNA被符号化，解码为设计的主要文脉，贯穿整个空间的始终。

DNA解码为接待大厅形象墙上的艺术装置——业主与其客户的LOGO相濡以沫。

DNA解码为柔美的照明装置。

DNA解码为方便务实的办公操作台，一个办公组团连接另一个办公组团，将窄长的办公空间很好地组合为一个有机的整体。

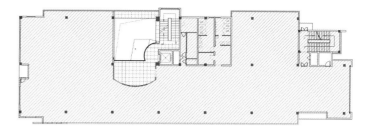

平面布置图一

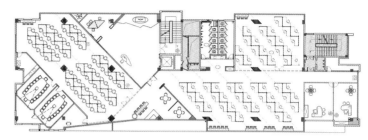

平面布置图二

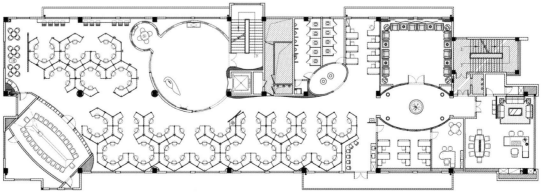

平面布置图三

Office of ZhengMao Photoelectric Co.,Ltd, Shenzhen
深圳正茂光电办公室

设计师 | 王五平　　设计单位 | 深圳五平设计机构　　项目地点 | 中国深圳　　建筑面积 | 2000平方米
主要材料 | 红砖、环氧地坪漆、鹅卵石、多乐士乳胶漆、光纤灯、铝方通

Viewing the project, the coarse cobblestone, the natural red brick wall, the clean epoxy floor coating and the naked ceiling conduits are emphasizing one of the pursues the project has, whom is named as environmental protection. The conception of environmental protection, altogether with the inventive ideas, is completely showed through this project.

In the reception hall on Floor 1, the function here is no more than the exhibition and reception. The ceiling used optical fibre lighting decoration. The upper parts of the walls at two sides were designed with irregular light holes. The regular arrangement was only to show another side of the light's beauty. The hollowed utter parts of the wall are in harmony with the red brick wall on Floor 2, so the felling of origin can be created. The things with the atmosphere the wall lamps create, give the whole area a deep meanings. In addition, the designer used Logo of the company's on the background to make a concrete meaning for the project.

The project has a scenery under the stair. The ceiling used red style decoration and in the utter part the designer has properly made use of the relations among marbles, cobbles and woods to add funs to the area.

Floor 2 is of the most importance to the project. Except to the reasonable arrangement for functions, the designer were also using innovative methods. For example, The reception area at the stair was designed to be 2 circles that can be formed as 8. What is more, the ceiling could be matched with 8. Therefore, the feeling of the style here is very vivid. The opening office area was divided into several independent spaces by the curtains. Meanwhile, the visual effect of the columns are also weakened.

The corridor design of the meeting room is also an excellent work. The existed columns were painted in white, being at contrast with the red brick wall. The shapes of the wall, altogether with the lights tossed by the wall lamps, give the area a sense of beauty as well as that of infiltration.

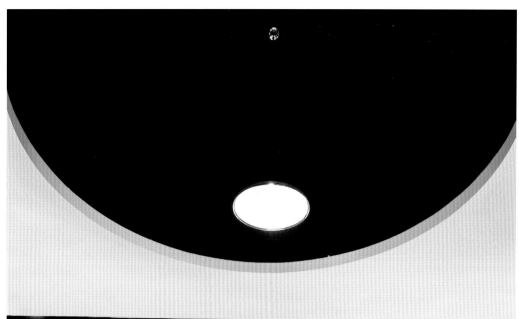

粗犷的鹅卵石、天然的红砖墙、洁净的环氧地坪漆、裸露的天花管道，这一切无不在强调着本案的一个诉求：环保。环保的设计理念且不失创意的思路，在本案也彰显得淋漓尽致。

在一楼的接待厅里，没有规划太多的功能空间，只为接待展示而用。天花采用光纤灯装饰，两侧墙体上部设计成不规则的方形透光孔，有序的排列组合，只为表现光还有另外一种美。中空下墙体部分则呼应二楼红砖墙饰面，以表达原汁原味的感觉，加上壁灯氛围的营造，整个空间意境深远。背景则采用公司logo元素，简洁而富有意义。

楼梯下面则设计了一个景观，天花采用红色的造型装饰，下方则采用大理石、鹅卵石、木板三种材料，丰富而又有情趣。

二楼是本案设计的重点，除了功能布局合理之外，在创新手法上也多变求新，如楼梯口和接洽区则设计成两个圆形，构成一个"8"字，天花与之呼应，这样一个抽象的"8"字在大厅中央，生动而意义深远。两侧开放式的办公区，通过屏风隔断设计形成相对独立的一个空间，同时也弱化了旁边几个大柱子的视觉效果。

会议室旁边过道设计也是本案设计的一个亮点，保持原有柱子刷成白色，和红砖墙形成鲜明的对比，凹凸有序，加上柱子的壁灯打出来的光束，极具空间渗透性和美感。

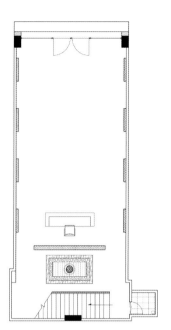

平面布置图一

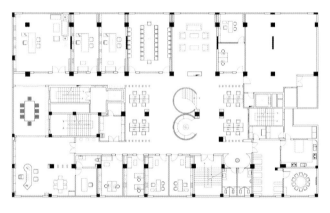

平面布置图二

Office at East Peace Road
和平东路办公室

设计师 | 许宏彰　　设计单位 | 德力设计　　项目地点 | 中国台北　　建筑面积 | 118平方米
主要材料 | 灰镜、烤漆玻璃、铁件、栓木、意大利压纹砖、意大利木纹砖、烟熏橡木、印度黑大理石、进口马赛克

After a long term preparation, The area of Office Peace's, whom was designed by DLdesign, is getting larger with better sample& materials room. This change can also give convenience for designer's choosing on materials. The brief meeting room was moved to the open dining room and mid-island kitchen. Therefore, the area can supply the clients a comfortable living environment. Lishui Office at the original site is transformed to DLliving. So in there it can supply more services for furniture customization, accessories matching and purchasing. To create an environment that can be stay away from interferences, DLlving would only supply the service for customization at the beginning.

The entire color of the Office Peace is quite and low-profiled. Whatever the floor or the matching of the accessories were used for creating a quite area in peace. At the same time, the designer had successfully made use of the LED lamps to create the visual focus points for each area distinguished in different functions. With all of these, everyone entering this area can deeply feel the power to absorb the interferences of emotional and out-environment. To take the bed temper's place, the building supply people a quite and elegant area for their further exchanges on life.

奉茶幕后团队"德力设计"筹备多时的新址"和平办公室"空间更宽阔，而且有了更完善的样品材料室，方便设计师选搭建材，简报会议室则移往开放式的餐厅与中岛厨房，希望借由场域的空间暗示，提供给客户一个更惬意、舒适的居家环境。至于，旧址"丽水办公室"将转型为"德力家饰"，提供更完整的家具定制、家饰搭配、家饰选购的服务。为了塑造一个不受干扰的环境，初期"德力家饰"将采用"预约制"为大家服务。

"德力设计"的和平办公室的整体色系沉静而低调，不论是地板柜体或是家饰配搭，都是希望创造一个低干扰的静谧场域，同时运用LED聚焦灯让不同机能设定的空间各具视觉焦点，让所有一踏入场域的人都可远离外界纷扰，取而代之的是一个"静谧"而"优雅"的场域，打造出一个让人安住放心和进一步交流设计生活品位的空间。

平面布置图

Vinistyle Cosmeytic Office Building of Sumei Group
苏美集团vinistyle化妆品办公楼

设计师 | 沈烤华、崔巍　　设计单位 | 南京沈烤华室内设计工作室　　项目地点 | 中国南京　　建筑面积 | 600平方米
主要材料 | 塑胶地板、软膜、橡木地板、欧林办公家具　　摄影 | 裴宁摄影

As a cosmetic training base and office, this case reflects its industrial personality everywhere. Although it's a base and office, the decoration is not made in totally soft and beautiful colors, but by neutral warm colors to create a cozy but serious office. The style is modern simplicity, and architecture frame is made of geometry and straight lines. Then the feelings of personalized, fashionable and modern city come into being. The height is made good use of, and each functional unit is connected by corridors, the space presents dream-like atmosphere by adding the floor material and unique ceiling with lightening effect.

The style of every functional area is specially designed and united in a whole. Different atmosphere is created in colors according to different areas. The office is fashionable, reasonable functional and cozy environmental. Economical but practical materials are used, and different colored plastic floor is adopted by considering the function of each area. Irregular cylinder soft filmed lightening structure on the ceiling create artistic atmosphere in the space.

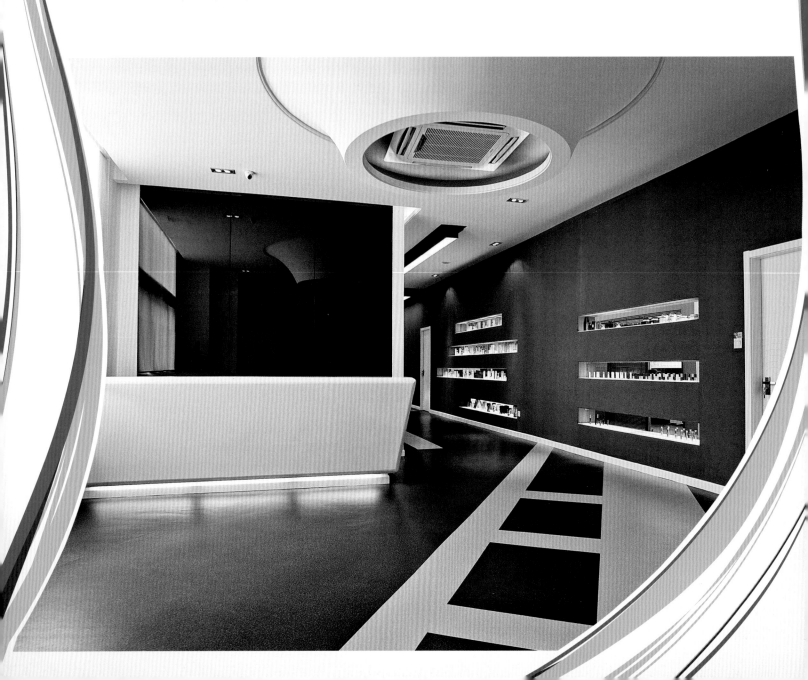

本案是一个化妆品培训基地和办公楼，从行业特点出发，空间随处体现十足的个性感，虽是化妆品培训基地及办公空间，却并没有用完全柔美、靓丽的色彩来打造整个空间，而是采用了一些中性、温婉的色彩勾勒出一个温馨而又不失严谨的办公场所。风格主要体现现代主义的简洁，以几何、直线构成建筑框架，打造出个性、时尚、都市感。设计中充分利用了层高优势，以走廊连接各个功能单元，结合地面材质区分和独特的吊顶及灯光效果，使空间形成了若即若离的效果。

各个功能区的风格都独具匠心，而整体又非常和谐统一，根据不同的场所利用色彩营造出不同的氛围，追求时尚，创造一个功能合理、简洁明快、环境优美舒适的办公场所。使用了经济实用的材质，结合各区域的功能采用了不同颜色塑胶地板铺设，顶部以不规则柱形的软膜灯光造型给空间营造出艺术气息。

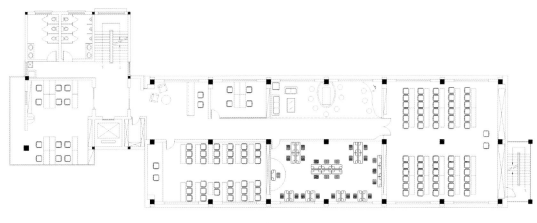

平面布置图

Disonna's Factory Reconstruction and Indoor Design
迪桑娜厂房改造和室内设计

设计师丨王文亮　　设计单位丨汉诺森设计机构　　项目地点丨广东深圳　　建筑面积丨外立面728平方米　内部面积4210平方米
主要材料丨钢架结构、木丝水泥板　　摄影丨汉诺森设计机构

This is a typical modernism design. However, it can't be acknowledged as simple. In fact, it is simple only referring to the forms and means, including people's pursues for sense and strengths. This is a tactic and explain on the visual effects and emotional appeals.

The designer used a quantity of all-steel structures in the erect outside of the construction. The sole middle gray steel, forming the vertical element, bring the outlooking of the building a magnificent sense. The hollowed space formed at the steel structure and original erected outside was used to be a expansible motive space by us. The step stair between floors supplies people the extra funs of outside activity's on one hand and an assistance on emergency evacuation on other hand. The most important thing is the humanism erected outside had weakened the rigid and out-of-fashion form of the building, and reduced the sunlights and temperatures. So the consumption on electricity of the air conditioners can be reduced either.

On the right side of the stair, we built an affiliated building to increase the using area of the floor and break down the rigidity of the large-size wall and import the lights.

While enter the reception area on Floor 1, the generation technique of the calculator help us complete the design for irregular

theme wall. Being compared with the symmetrical reception areas at two sides of the front table, the wall had subtly imported the lights through the ceiling designing. We chose environment-friendly remade wood cement boards to make the wall. Therefore, we can reduce the harms created by the wood floor and fibre carpets to the largest extent.

We also adopted a constructive materials on both outside and inside of the buildings. The materials, named wood cement board, are made by cements and woods and are as lighter as the wood. It also owns the characteristic of insulation and elastic. The materials are as firm as the cement, so it can meet the anti-fire, anti-rust, moisture-proof need of the Southern China's climate.

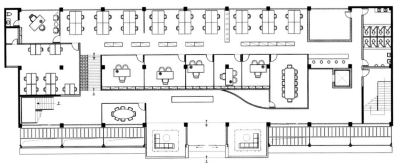

平面布置图

这是一个典型的现代主义的设计，但不能单纯理解为简约，事实上，这种有严格意义上的形式和手段的简约，包含了人对知觉、力量的追求，是一种策略，是对建筑赋予视觉效果和感情诉求的诠释。

建筑外立面采用大量的全钢架结构，此唯一的中灰色钢构横向元素，从视觉上带来有完整体量的宏大外观。在钢架与原外立面形成的中空空间，我们大胆地设计成延展式活动空间，楼层之间的踏步楼梯的设置，既提供了人们意外获得的户外活

动乐趣，又可在人员疏散方面起到安全辅助作用。更重要的是，人性化的外立面设计弱化了原建筑物呆板的陈旧形式，并在中国南部酷热气候下起到了一定的阳光阻隔作用，从而降低空调能耗。

在楼的右侧，我们增建了附楼，增加了平层的使用面积，破除了大面积墙面的沉闷，引入了光线。

进入一楼的接待区域，计算机生成技术帮助我们完成了不规则主题墙面的设计完成，相对前台两边对称的接待区，依靠巧妙的天花设计，引进了光线。内部我们选用了环保再造型材木丝水泥板铺设墙面，最大程度降低由木质地板及纤维地毯带来的环境负担。

楼体内外部采用了同一种建筑型材料：木丝水泥板，其由水泥和木材混合而成。如木材般质轻，具有弹性和隔热性能，也如水泥般坚固，满足南方防火、防潮、防霉的气候需求。

Kun'en Investment
坤恩投资

设计师丨蔡进盛　设计单位丨方块空间　项目地点丨中国南昌　建筑面积丨546平方米
主要材料丨石材、墙纸、乳胶漆、实木地板

Nowadays, the atmosphere and aesthetics become the focus points for office designing. Under a good environment would the employees's mood be good, and the working effectiveness would spontaneously be increased.

The project mainly emphasizes the importance of quality and taste. Simpleness and greatness are the foundation for the designs, also bestow the enterprise an elegant quality. For these, an effective, swift and humanistic working environment comes into being.

The characteristics of the project is no other than subtly combining the style of China and that of contemporary together. The simplified designs of the contemporary style would been seen everywhere in the hall and lobby. When turns to the office and meeting room, the viewer would feel strong China Style instead. Though there are two completely different styles, in this project they are not contradict to each other since that the designer generously connects them smoothly and naturally through a subtle method to let the viewer accept the conception of integrity while feel the two kinds of atmosphere.

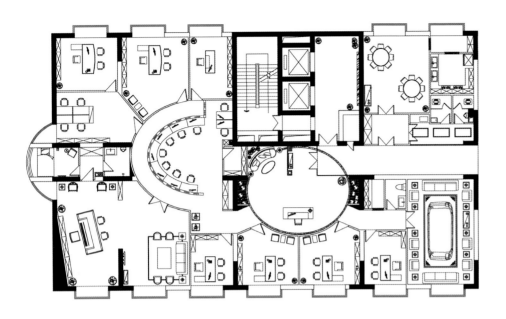

平面布置图

如今办公场所也讲究氛围和美观了，在美好的环境下办公，人的心情也会变得舒畅，办事效率自然也会提高。

本案总体设计强调品质与品位的凸显，简单大气是整个设计的基调，并赋予公司一定的内涵，打造高效、明快、人性的工作环境。

该项目的特点在于现代风格与中式风格的巧妙结合。大厅与大堂都是采用现代时尚简约手法，转到办公室、会客厅时，人们又可以感受到浓浓的中式氛围。虽是两种截然不同的风格，在此却并不显得突兀，设计师以巧妙的设计手法使这两者间衔接得非常自然，让人们在感受两种氛围的时候，同样能有整体性的概念。

Santillana
桑提亚拉

设计师 | Juan Carlos Baumgartner　　项目地点 | 墨西哥　　摄影 | Paul Czitrom

One of the most important publishing groups internationally needed to change offices to meet its needs for space. We made the most of the move to also make a much-needed culture change.

Its offices had no identity and did not get any benefit from the great object they sell. For this reason our design is based on the Book, making it the great theme of the office. The reception revolves around 11,000 irregularly arranged books, cascading from the ceiling until they become the main wall. A gray Corian wall ends up embracing this sculpture and the company logo, subtly delineated with a router, allows the book to take center stage.

As we enter the floors there are small lobbies as a prologue to what takes place inside. Informal meeting spaces and phone booths to receive suppliers without having to go through the whole office. Next, an open space with infinite possibilities. Each brand has its own personality, colors on columns, ceilings and furnishings. Each as a different chapter. Private offices attached to the core of the building free up the facades and allow free passage of natural light to all employees. Casual collisions are scattered on all floors in strategic locations, bringing our tour to a finale.

The overall image creates a hybrid between minimalist architecture

and the eclecticism of the book. Pure materials and colors throughout the spaces and furnishings let the book be, in all its chaos, lending touches of color.

During your visit you immerse yourself in the space, just as you would a good book.

一家在世界上占有重要位置的国际出版商为了他们的空间需求而重新要求改变办公楼。我们做的则是推进这一步，并同时实施亟待的文化改变。

办公楼没有任何身份，也不能从他们所出售的产品上获得任何利益。因此，我们将区域的主题建立在"书"上：整个接待区域被大约11000本不规则摆放的书籍环绕，这些书籍自天花板呈瀑布形摆列下降，最终成为了一道主墙。一堵灰色的可丽耐墙基装饰着雕塑与公司标识，并在周围巧妙地雕刻了一道刨槽，这样使得书籍能够占据主要舞台。

当我们进入房间，这里有一个小小的回廊作为进入内部空间的序曲。非正式的会面区和电话亭使人们不进入办公区就能获得供给品。接下来，就是一个拥有无限可能的开放空间。每一个品牌都有它独到之处，柱子、天花板和装饰的主色，就如每一个篇章的不同一般。临近建筑物核心区域的私人办公室的对外面是可视的，使得自然光可以射入，雇员也能看见内部。在战略位置的地面上夹杂偶然的强烈对比让这一次旅程达到了重点。

整个景象创造出了一个极简的建筑与书籍的折中主义的混合体。充斥整个空间与装饰的单纯材料、颜色使处于嘈杂环境的书籍始终保有颜色的主导位置。

在整个旅程里，你沉浸在空间当中，就如沉浸于一本好书里。

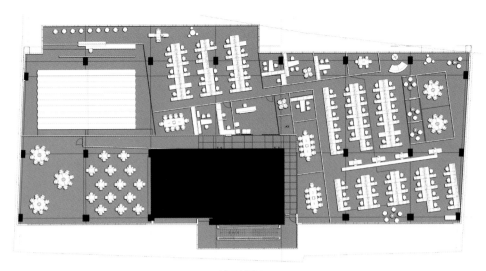

平面布置图

Cabel Industry
软件公司Cabel Industry

设计单位 | Massimo Mariani   设计团队 | Elda Bellone、Roseda Gentile、Alessandro Mariani、Giovanni Lunari、Simona Baronti
项目地点 | 意大利恩波利   建筑面积 | 4500 平方米   摄影 | Alessandro Ciampi

Just out the edge of the town of Empoli, the building is the Cabel Industry headquarters (a company dealing in computer systems for banks), it covers an area of approximately 4.500 square metres over several levels and it is incorporated on the small local industrial estate setting up new dialectical relations with the local manufacturing fabric.

Commitment asked us to project an office building to be constructed using industrial methods by keeping down costs and time of construction. We decided to design a new precast concrete panel featured to be stand on the two main facades.

Partially set into the ground, the building is composed of two extended floors out the ground level and a vault under. Along the main front the facade is protected by a long strip of public landscaping running parallel to the road.

The visitor arrives on the ground level through three suspended bridges launched on a large excavation which lights the vault designed to house expositions and art installations. At night time this empty space becomes a pool of light which allows the architecture to detach itself from the ground, making it look like a lightweight multi-colored object suspended in the dark.

Inside the building the underground level holds a printing facility

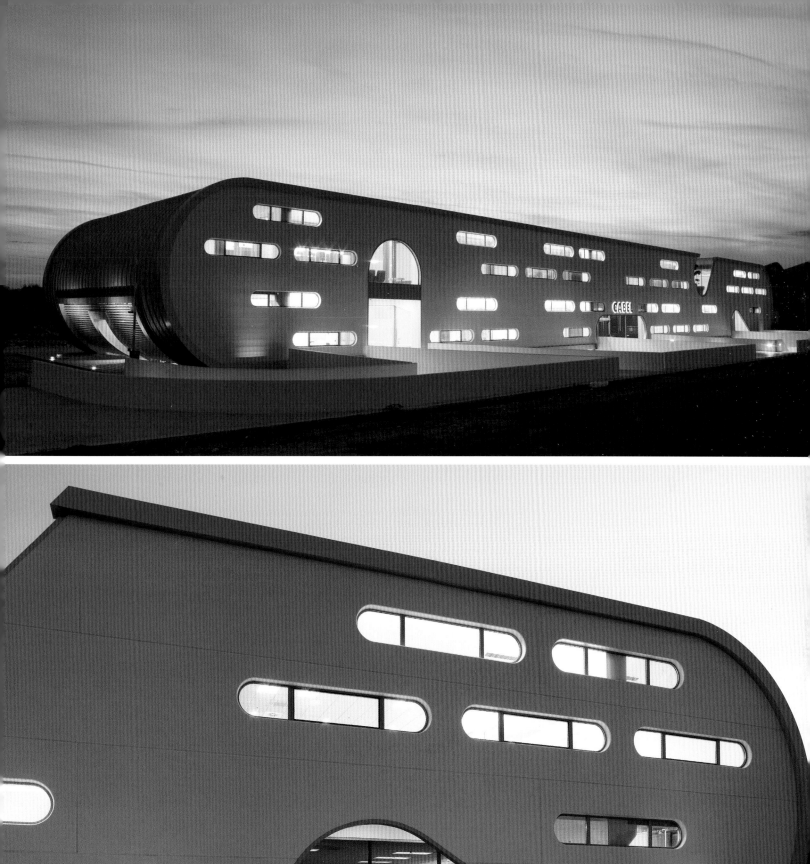

and other spaces allocated for various different kinds of activities; the ground floor features all different sorts of glazed spaces, ranging from open plan spaces to small isolated cells, in a modern interaction of liquid working areas like acquarium. In contrast the first floor accomodates the management space, with a small internal patio and terrace in-between them. The elevators and distributional stairs take up the middle section of the structural grid, made of stone tiles with steel elevators.

The gently curving building shell, windows, entrance gaps and furniture decoration all feature the same basic stylistic design.

During the day coloured glass create liquid chromatic effects inside black and white offices, instead of night time when coloured cuts project out vivid lighting effects underlining holes, cuts and shapes of the building.

The building is constructed out of prefabricated concrete elements (painted with white protective enamel) and its roof and end sections are tiled in grey aluminium. Starting from outside, external wall consists of precast concrete pannel, 80 mm thermal insulation, Knauf plasterboard. Everywhere suspended floors are made of fine porcelain (gres). Inside ceilings consist of plasterboard (modular or continuous) with thermal insulation under. So the thermal factor among the floors is optimized for reducing energetic consumption in climatic control.

The entire roof is covered with a system of photovoltaic solar panels using amorphous polycrystalline tecnology, carefully positioned so as not be visible. Thanks to this system, which is capable of generating approximately 150 kW, the building is almost totally self-sufficient from an energy viewpoint.

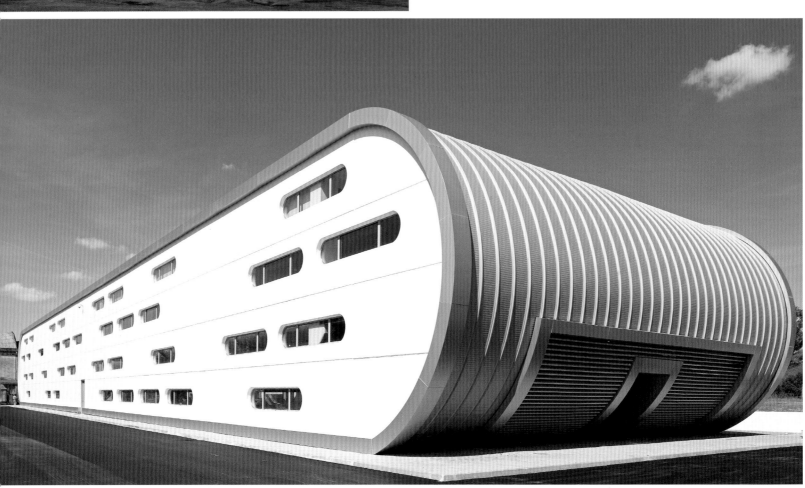

软件公司Cabel Industry（为银行研发计算机系统）的总部办公楼坐落于意大利恩波利，占地约4500平方米，由多层构成。它是由当地多家小产业公司合并而成，与当地的面料生产商形成了一种新的辩证关系。

鉴于客户要求用工业的方法压缩成本和施工时间来完成这一新建项目，我们决定在两个主立面设计一个新的混凝土预制板。部分插入地面，大楼由两个延升楼面和拱顶组成。主立面前是与道路平行设置的公共美化带。

游客可以通过三座延伸桥到达地面一层，桥下是用于展览及陈列艺术品的地下空间前的绿地，可通过阶梯到达。晚上，该空间洒满了灯光，使该大楼截然而立，犹如黑暗中闪烁淡淡光芒的多色体。

地下一层主要用于展览、陈列艺术品以及安排各种活动，地上一层设置有各种功能的空间，从敞开式空间到孤立小单元等，像水族馆内液体装卸区的现代互动。地上二层则与之相反，该层主要用于管理功能，有一个小型内部天井及平台。电梯和楼梯占据空间中间部分，由石材瓷砖制成加钢制电梯。

微微弯曲的建筑外壳、窗、入口区和家具装饰都采用基本相同的风格设计。

白天，彩色玻璃为黑色和白色的办公室带来液体色效果，然而到了晚上，彩色会削减生动的照明效果，取而代之的是强调建筑外形效果。

该建筑物采用预制混凝土构件建造（白色防护珐琅绘），屋顶及建筑体两端包以灰色铝制外层。从外及里，外墙都是由预制混凝土板、80毫米保温层、可耐福石膏板组成。悬浮地板是由精美瓷器制成（GRES）。天花板内层由具有保温作用石膏板构成。因此，最大化地保证了楼层间的保温效果，减少气候变化带来的能量消耗。

建筑屋顶上覆盖有采用无定形多晶体技术制造的光电太阳板，太阳板的布置经过了精心设计，所以不容易看到。这些太阳板能够产生150KW的电力，基本能满足建筑对能源的需求。

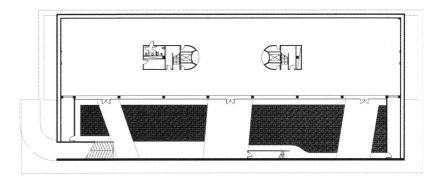

平面布置图一

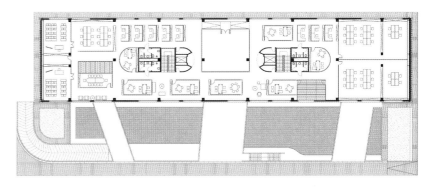

平面布置图二

立面图一

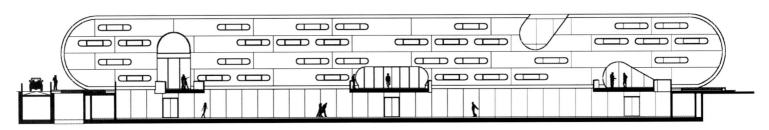

立面图二

Created Office
克里德办公室

设计师 | 张德良、殷崇渊　设计单位 | 演拓设计（台北 东莞 佛山）　项目地点 | 中国台北　建筑面积 | 688平方米
主要材料 | 大理石、抛光砖、木纹PVC、人工草皮、烤漆玻璃、人造石、毛丝面不锈钢、皮革、地毯、茶镜　摄影 | 游宏祥

The story line of this case is the image of cloud which has been reflected in the whole office. Pure and warm colors and fully expressed Persian carpet present contented stable picture in the light. Open and thorough meeting area creates spacious feeling which can extend the vision to the most the first time, afterwards, transfer to next section with the flow of cloud.

The background wall of public leisure area is made of artificial turf, bringing natural tranquility to busy work.

Create an open and elastic space, match suitable moving lines, increase interaction among people, and then a real LOHAS working area will come into being.

本案的设计以云的意象为主,将其化做具体表现于整体的办公空间,色彩基调单纯温润;满铺的波斯地毯,透过光的洗礼,散发出内敛的气息,通透开阔的会议区,第一时间奠定了足以延伸视野的开放感,再随着云的流动转向下一个场域。

在公共休憩区域背墙以人工草皮铺陈,为繁忙工作的人们带来一丝宁静。

创造、开放而富有弹性的空间,搭配合宜的动线,增加人们彼此间的互动,打造一处真正的乐活工作区域。

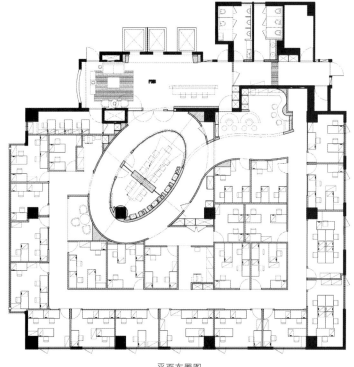

平面布置图

Shanghai Oulin Office
上海欧林办公室

设计师 | 潘均　　设计单位 | 尚辰设计　　项目地点 | 中国上海

This case is a design for office with the modern style. The design divided the space into working area and meeting area according to principle of dynamic and static. The entire space is based on the simplified and modernized designing method. Whatever is the functional combination or the use of the materials, or the matching of colors, are all showing the feeling of modern, practical as well as of the simple. Not only had it satisfied the needs for working, but also can show the modernization and anesthetic of the decoration. White is the main color for the space, and the swift red and vivid green are also used here to emphasize the modern feeling for the office. To create the steady and serious atmosphere for the office, the designer used the deep-colored wood boards to make the background wall in the reception area. The round troffer in the ceiling of the hall made the space more vivid. In here we can deeply feel the ease hand harmony of the humanized design of this office space.

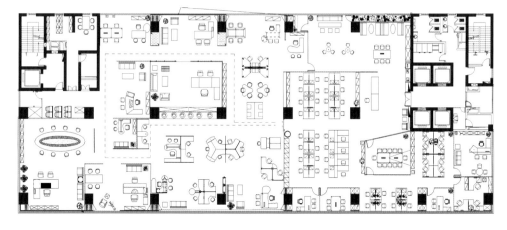

平面布置图

　　本案例为办公室设计，设计风格以现代风格为主，设计根据动静分区的原则对空间进工作区域及会议区划分。整个空间以简约、现代的设计手法为依据，无论功能组合，材料运用，以及色彩搭配，都体现了现代，实用，简洁，明了，即能满足办公需求，又能体现装饰的现代与美感。白色是空间的主基调，明快的红与清新的绿不时穿插其中，强调办公的现代感。前台接待区的背景墙以深色木板拼贴而成，营造了办公室沉稳、庄重的氛围。大厅的天花上，一个个圆形隐藏式的灯槽，给空间点缀得生动而富有灵性。在这里可以感受一份轻松和谐的心境，从而传达办公空间的人性化设计。

图书在版编目(CIP)数据

源动力：新锐办公空间／未来文化主编．—武汉：华中科技大学出版社，2013.3
ISBN 978-7-5609-8434-6

Ⅰ.①源… Ⅱ.①未… Ⅲ.①办公室－室内装饰设计－作品集－世界 Ⅳ.①TU243

中国版本图书馆CIP数据核字(2012)第242212号

源动力—新锐办公空间

未来文化 主编

出版发行：华中科技大学出版社（中国·武汉）	
地　　址：武汉市武昌珞喻路1037号（邮编：430074）	
出 版 人：阮海洪	
责任编辑：王晓甲	责任监印：张贵君
责任校对：周　娟	版式设计：彭种玉

印　　刷：利丰雅高印刷（深圳）有限公司
开　　本：965mm×1270mm 1/16
印　　张：19
字　　数：280千字
版　　次：2013年3月第1版 第1次印刷
定　　价：298.00元 (USD 59.99)

本书若有印装质量问题，请向出版社营销中心调换
全国免费服务热线：400-6679-118 竭诚为您服务
版权所有　侵权必究